The Climate Swerve

ALSO BY ROBERT JAY LIFTON

Witness to an Extreme Century: A Memoir

Death in Life: Survivors of Hiroshima

Hiroshima in America: Fifty Years of Denial (with Greg Mitchell)

The Nazi Doctors: Medical Killing and the Psychology of Genocide

The Genocidal Mentality: Nazi Holocaust and Nuclear Threat
(with Eric Markusen)

Indefensible Weapons: The Political and Psychological Case
against Nuclearism (with Richard Falk)

Destroying the World to Save It: Aum Shinrikyo,
Apocalyptic Violence, and the New Global Terrorism

Home from the War: Vietnam Veterans—
Neither Victims nor Executioners

Thought Reform and the Psychology of Totalism:
A Study of "Brainwashing in China"

Revolutionary Immortality: Mao Tse-tung
and the Chinese Cultural Revolution

The Protean Self: Human Resilience in an Age of Fragmentation

The Broken Connection: On Death and the Continuity of Life

The Future of Immortality: And Other Essays for a Nuclear Age

Birds (humorous cartoons)

The Climate Swerve

Reflections on Mind, Hope, and Survival

Robert Jay Lifton

THE NEW PRESS

25 YEARS

NEW YORK
LONDON

Requests for permission to reproduce selections from this book should be mailed
to: Permissions Department, The New Press, 120 Wall Street, 31st floor, New York,
NY 10005.

Published in the United States by The New Press, New York, 2017
Distributed by Perseus Distribution

ISBN 978-1-62097-347-9 (hc)
ISBN 978-1-62097-348-6 (e-book)
CIP data is available

The New Press publishes books that promote and enrich public discussion and
understanding of the issues vital to our democracy and to a more equitable world.
These books are made possible by the enthusiasm of our readers; the support of
a committed group of donors, large and small; the collaboration of our many
partners in the independent media and the not-for-profit sector; booksellers, who
often hand-sell New Press books; librarians; and above all by our authors.

www.thenewpress.com

Composition by dix!
This book was set in Garamond Premier Pro

Printed in the United States of America

10 9 8 7 6 5 4 3 2 1

For Nancy Rosenblum
with love and astonishment

In a dark time, the eye begins to see.

—Theodore Roethke

Contents

Preface

This is a book about climate change but not a detailed study of its ubiquitous effects, or of its political requirements. Rather it is an exploration of mind and habitat, a meditation on what I call the *climate swerve*, our evolving awareness of our predicament. The climate swerve creates a mind-set capable of constructive action, and is a significant source of hope.

During sixty years of thinking and writing about nuclear weapons I frequently came upon people and events pivotal to the climate story. But it took me a long time to grasp the significance of these encounters and turn my focus to mind-sets related to global warming. I am part of that climate swerve.

Living in Hiroshima for six months in 1962 and interviewing survivors there gave me a hands-on sense of the human effects of nuclear weapons. Like global warming, those weapons raised doubts about the future of our species. I came to recognize that one can learn much by comparing these two apocalyptic twins and our responses to them. I came to see parallels, differences, and overlappings.

When talking to people in Hiroshima I found myself wondering about what went on, psychologically and politically, at the other end of the weapon. This concern with both survivors and scientists, with their voices and mindsets, will be prominent throughout this book. Survivors, like those in Hiroshima, seek meaning from their death encounter, and they may embark on a mission of witness that can be illuminating to their fellow human beings. Scientists created nuclear weapons and then warned us about their danger; it was scientists who revealed to us the devastating consequences of global warming. Scientists in particular, and professionals in general, bring special authority to nuclear and climate matters, which is why the few among them who become associated with various forms of denial, falsification, and rejection have done such great harm.

My work on climate change, then, owes much to my study over the decades of the mind's relationship to murderous nuclear devices and the world-ending threat they pose. But other studies I have done are also relevant to my approach to the climate story: my work on Chinese thought reform and the mind's response to totalism; Vietnam veterans and the American confrontation with atrocity; and Nazi doctors and the potential of professionals for active collusion in genocide. The destructive forces encountered in these studies shed light on human-caused climate disaster.

At the same time, the Paris Climate Conference in December 2015 was a stunning demonstration of universal awareness of the danger of global warming, of what I call the climate swerve. Virtually every nation in the world joined in a recognition that we are part of a single species in deep trouble. More recently many of the signatories have expressed further commitment to specific promises to reduce carbon emissions. This display of species awareness is unprecedented, and holds out the possibility that we humans may extricate ourselves from extreme climate catastrophe.

To be sure, there is profound resistance to such awareness, even at times a vicious backlash. The problem is magnified by right-wing political forces, sometimes termed "populist" or "ethnonationalist" movements, gaining power in parts of the world, groups that reject climate change and slander its advocates. Most notable here, and most dangerous to the world, is Donald Trump, the recently elected American president. He and people he has chosen to serve in his administration view climate change as nonexistent, not human caused, or a hoax. But even Trump and his allies cannot fully avoid, in some part of their minds, the recognition that harmful, human-caused climate change does indeed exist. In that sense they are less climate deniers than *climate rejecters*. Their danger lies in both the actions they resist and those they take. Yet in ways I will consider further, they too are susceptible to the forces at play in the climate swerve.

Climate change confronts us with the most demanding and unique psychological task ever faced by humankind. Yet we have the capacity to apply our minds to this task in the service of protecting our habitat and surviving—perhaps renewing ourselves—as a species.

No one claiming to be an intellectual or a concerned citizen can avoid confronting either the nuclear or the climate threat. But because the climate problem is all-enveloping, no individual person can adequately encompass it. My portion of it—the task I have set for myself—is the comparison of nuclear and climate threats as a way of focusing on the dilemmas we face in connection with our own prior and contemporary actions. I make no claim that this perspective will in itself decrease our carbon emissions or stem the overall rise in sea levels. But even as it enables us to take another look at the still-pervasive nuclear danger, I believe it can offer a measure of insight into grasping our climate menace and acting on it.

I

The Ultimate Absurdity

I wrote this book just after celebrating my ninetieth birth-day. I don't believe most reasons people give for their lon-gevity, but if asked about mine I would, with tongue only slightly in cheek, point to a longstanding sense of absur-dity. This does not mean that I belong to a philosophical or artistic movement that focuses on meaninglessness be-cause we know that we all will die or that that we inhabit an "indifferent universe." In fact I understand us to possess minds that can create meaning in our lives and find ways to preserve the piece of our universe that we call our planetary habitat. Rather, my sense of absurdity has to do with how much that all of us observe and experience is contradictory, less than rational, based on distorted fantasy.

To be sure, this sense of absurdity has considerable rela-tionship to the extremely destructive, and self-destructive, events I have studied: systematic efforts to control the human mind; the dropping of atomic bombs on populated cities; the pursuit of atrocity-producing wars; and physicians' vast-scale reversal of healing and killing.

But my sense of absurdity surely preceded these studies and has had more to do with everyday human foibles, very much including my own. I have long expressed my sense of absurdity in bird cartoons. The birds are stick figures—I have no artistic talent—but they can expose just about any kind of pompousness or dubious claim, including the very claim of the uniqueness of the individual self. In one cartoon, which I like to refer to as my existential classic, a small, naïve, enthusiastic bird declares, "All of a sudden I had this wonderful feeling, 'I am me!'" And a larger, more skeptical—even jaundiced—bird looks down at him and responds, "You were wrong."

This sense of absurdity has not prevented me from joining struggles to oppose what I take to be dangerous forces, especially those related to issues of war and peace and nuclear weaponry. To the contrary, I believe that it has contributed to my judgments, however fallible, about what to oppose and how to oppose it. It may even be a source of hope: what is identifiably absurd can and should be changed to something less absurd and more life-enhancing.

But the subject of global warming is absurd in a newly bizarre way. Its ultimate absurdity is this: by merely continuing with our present practices and routines, we human beings will increasingly harm our own habitat, the portion of nature we require to survive, and ultimately destroy our own civilization. We needn't start a war or make use of nuclear weapons. We needn't do anything—other than what we are already doing—to endanger the future of our species.

The "indifferent universe" that the philosophers of absurdity have been fond of referring to has indeed been indifferent to our actions. But those philosophers could not have imagined our darker, species-created climate absurdity.

The crucial interaction at issue is that of mind and habitat. Mind is our precious human achievement, a wonderful instrument that can do dreadful things. Our symbolizing minds cause us to reconstruct, with our cerebral cortex, every perception we have. We cannot see, hear, feel, smell, or touch without bringing to the experience a measure of our imagination. That imagination could create a Mozart horn concerto as well as a theory that the Nordic race can best be strengthened by killing all Jews. Yet our mind is our only means of addressing the physics and politics of global warming.

Our habitat is the portion or portions of the earth and its atmosphere—of "nature"—that we humans require to survive. While the word "habitat" suggests specific geographical places, the human genius for adaptation has expanded our habitat to include our entire planet. As the lyrical anthropologist Loren Eiseley wrote: "The creature who could clothe himself in fur or take it off at will, who could, by extension of himself into machines, fly, swim or roll at incredible speeds, had simultaneously mastered all of earth's environments with the same physical body" so that "no longer could man be trapped in a single skin, a single climate, a single continent, or even a single culture. He has become ubiquitous." But with this omnivorous adaptation and unlimited symbolizing imagination, man "has the capacity to veer with every wind, or stubbornly, to insert himself into some fantastically elaborated and irrational social institution only to perish with it." In that way man became "the single lethal factor at the root of declining or lost civilizations."

Facing climate threat, our efforts at adaptation have included reckless consumption of the planet's energy resources, notably its fossil fuels, while numbing ourselves to the recognized consequences. At the same time we have succumbed to the power of corporations and nation states at the forefront of this rapacious quest.

No wonder some psychologists and neuroscientists look for an explanation of this behavior in the wiring of our brains. They claim that the brain enables us to deal with immediate threat but not with prospective possibilities. I think that view is half true, and highly misleading. Our brain wiring matters a great deal, and we are surely better at taking in a direct experience than at imagining events in the future. But at the same time a distinguishing evolutionary characteristic of what has come to constitute the human mind is precisely our capacity to imagine what has not yet happened.

With global warming, moreover, the catastrophic future is increasingly visible in a disaster-dominated present: in the high temperatures, droughts, extreme fires, and coastal flooding now occurring throughout the world.

The climate swerve is a manifestation of collective imagination, prospective and immediate. The chapters that follow explore these efforts to pit the human mind against the ultimate absurdity of global warming—to make use of our mental resources, including our political and social capacities, to confront this absurdity and find our way toward climate sanity.

Working now from what Theodor Adorno and Edward Said would call my "late style," I have already witnessed a dreadful interaction of mind and habitat in connection with nuclear weapons, and with Hiroshima in particular.

2

Hiroshima as Pollution

From my interviews in Hiroshima I learned of rumors that circulated immediately after the atomic bomb struck, rumors that revealed survivors' anxious sense of the vulnerability of their habitat. The most persistent of these rumors, and for many the most disturbing, was that trees, grass, and flowers would never again grow in Hiroshima. Because of the bomb's "poison"—its radiation effects—the city would be unable to sustain vegetation of any kind. Nature would dry up altogether; life would be extinguished at its source. The rumor suggested a form of desolation that not only encompassed human death but went beyond it.

Another rumor was that for a period of seventy-five years—or perhaps forever—Hiroshima would be uninhabitable: no one would be able to live there. Suggested here was the idea that the city itself, the habitat of all who dwelled in it, was contaminated beyond the point of sustaining human life.

There was also the sense among survivors that the bomb had so altered the natural world that the Americans were capable of further altering it in any imaginable way. Hence rumors that America would mount new attacks with "poison gases" or "burning oil"; that America, having dropped such a terrible "hot bomb," would next drop a "cold bomb" or "ice bomb," which would simply freeze everything so that everyone would die; and even a rumor that America would drop rotten pigs with the result, as one survivor put it, that "everything on the earth would decay and go bad."

All of these rumors were related to observed radiation effects, acute and delayed, or what I came to call a fear of *invisible contamination*—the fear among survivors that the same mysterious, pervasive, and deadly poison that had entered their individual bodies would engulf their entire city.

As it turned out, the environment was not permanently destroyed. The appearance of cherry blossoms the following spring even conveyed a partial sense of renewal. That environmental recovery was in keeping with a very different view of nature expressed to me by a survivor who quoted an old Japanese (originally Chinese) saying: "The state may collapse, but the mountains and rivers remain." Here the message expressed was that whatever we humans destroy, nature will somehow endure. Though events seemed to confirm this second, more hopeful view, Hiroshima survivors retained feelings that they had come close to experiencing the destruction of everything, to experiencing, as several put it to me, "the end of the world." Indeed the desolation expressed by those early post-bomb Hiroshima rumors turned out to anticipate later responses of people everywhere to the interactions of mind and habitat.

Overall, people in Hiroshima became deeply uncertain about how much one could depend upon the natural world to keep human beings alive. And many people elsewhere, having heard about or encountered images of Hiroshima and the weapon that destroyed it, took on a similar sense of cosmic uncertainty that continues to haunt human beings everywhere. We could not confidently depend upon "nature" to protect us from this new weapon. We learned that, whatever our destructive power, mountains and rivers may indeed remain.

What may not remain, however, are precisely the elements of nature necessary to human life: an atmosphere surrounding the earth that is not overheated, and oceans that are not rendered acidic and dangerous to the land around them. In other words, it is precisely the human habitat, and that of other plant and animal species—just a small part of nature — that is threatened. The rest of nature will be okay. If our bombs inundate the earth with deadly radiation, or bring about a nuclear winter by creating debris that blocks the rays of the sun, or if our carbon and methane emissions significantly increase climate temperature, then the prevailing mountains and rivers will do little to sustain the life of our species.

Hiroshima and the Ultimate Pollution

Early in my work I struggled with the relationship between nuclear threat and danger to our habitat. I completed my Hiroshima study in 1962, and a few years later I wrote most of a talk that I called "Hiroshima and the Ultimate Pollution." I apparently gave the talk on a couple of occasions but had forgotten about it until just a few years ago when my assistant rediscovered it among my papers deposited at the New York Public Library. In the talk I discussed those Hiroshima rumors and their suggestion of our new capacity to destroy, or come close to destroying, our natural world. I spoke of the "breakdown of ecological balance" and said that we "must use such terms as 'poison,' 'deterioration,' 'degeneration,' and 'starvation,'" as we are "speaking of nothing other than death, whether partial, symbolic, or total—and not only death but grotesque and premature death." I was talking about both nuclear weapons and extreme environmental pollution. And I went on to insist that "to preserve our planet and avoid these deadly breakdowns is to speak of life—not only in the individual person but the life of our species and that of other species as well."

I believed that those Hiroshima rumors were prophetic

for all of us, possessed a Cassandra-like quality, which we had better heed. The term "global warming" was not yet in general use, and I was clearly unaware of the important scientific findings about climate change that had already been recorded. But I was groping for a way of comparing, and finding parallels between, nuclear and climate threats.

The "ultimate pollution" I referred to—of individual bodies and the atmosphere outside of them—was created by an atomic bomb.

Ecocide

Around the same time, I and a loosely connected group of activists began to talk about "ecocide," a term we used to suggest large-scale destruction of the environment. Interestingly, we first spoke of ecocide not in relation to nuclear weapons or climate change but to the U.S. military's use of the herbicide called Agent Orange during the Vietnam War. Agent Orange was actually a combination of herbicides, named because of the signature orange stripe across the barrels in which it was stored; it was sprayed widely on vegetation in South Vietnam for the purpose of depriving guerrilla forces of protective terrain. Agent Orange contained benzoic acid and dioxin, and turned out to be extremely harmful not only to plant ecology but to people as well, increasing rates of cancer and causing birth defects, brain disorders, and other grave diseases in Vietnamese and Americans who were exposed to it.

The term "ecocide" was first used by a friend and colleague at Yale, Arthur Galston, whose relationship to Agent Orange was both tragic and inspiring. Tragic because Galston's work as a young botanist on a benzoic acid compound was utilized by American military scientists in producing Agent Orange. Inspiring because Galston was deeply troubled by the part his work played in extending war into environmental destruction, spoke often about his sense of guilt and responsibility, and became an extraordinarily articulate antiwar activist who made many trips to Vietnam and China, focusing always on the dangers of Agent Orange.

In that way he and other scientists eventually persuaded the American government to end the herbicidal project. I collaborated with him in a number of antiwar activities and observed the passionate intensity of his very personal involvement. His dedicated opposition to the use of the substance he had helped create, and his subsequent work in the new realm of bioethics, gave powerful expression to what I have come to call an *animating relation to guilt.* What I mean by that is the converting of self-condemnation into the anxiety of responsibility. That in turn can result in considerable achievement.

Another friend, Richard Falk, did much to bring the term ecocide into the international legal realm. Falk, an authority on international law, pointed out that American behavior in Vietnam "provided the first modern case where the environment was selected as a 'military' target appropriate for comprehensive and systematic destruction." He proposed a U.N. ecocide convention in 1973 that, although not officially accepted, had considerable influence over the years. His proposal was to become a basic document for a contemporary Ecocide Project, a widespread effort to render ecocide an international crime parallel to, and part of, the U.N. Convention on the Prevention and Punishment of the Crime of Genocide. Ecocide, then, has long been my concern, even if I took considerable time to realize its full significance. Without being aware of it, I was undergoing a change in consciousness in connection with the human habitat. That is, I was becoming part of the climate swerve.

Taking On (or Being Taken On by) Climate Change

Recently I have wondered at the hiatus between my work on Hiroshima and my concern about climate issues. I remember a conversation I had in 2013 with Jonathan Schell, the most influential of all writers on nuclear issues, about our having both been slow to take up the issue of climate change. We agreed that there was a deep connection between nuclear and climate threats and that climate was the all-enveloping one. Now I wonder whether both of us were not held back by a faith in the ultimate stability of nature, a faith that could cloud the minds of even those who were strongly aware of nuclear threat. I believe that Schell and I were no different from many others in assuming, however unconsciously, that those mountains and rivers would indeed outlast—and perhaps contain—the devastation of nuclear war.

In my case I had to be prodded by a friend, a member of the Wellfleet group on psychology and history, who pointed out in 2012 that we had never taken up the subject of climate threat. I had formed the group in 1966, together with Erik Erikson, and had hosted all of its annual meetings in my home in Wellfleet, Massachusetts. We had frequently discussed nuclear weapons, the Vietnam War, and psychological perspectives on various historical events. My friend suggested that we should connect climate change with our ongoing concerns, and that I in particular should have something to say about the issue, just as I had about nuclear threat. I experienced a direct and immediate affirmation, as one does to ideas that make contact with unrealized inclinations of one's own. Climate change became a central issue of the Wellfleet group, and I began a systematic comparison of climate and nuclear threats, drawing upon past nuclear explorations that I could now see in a somewhat different light.

In that way my study of climate change was anchored in previous work. In that earlier research I had made use of extensive psychological interviews with people who had undergone various forms of duress. For instance, my work in Hiroshima was primarily based on in-depth psychological interviews, supplemented with what I call a mosaic approach, an immersion in the historical and cultural forces providing the context for those interviews. But in this and other research, the interviews themselves had been basic to the method; through them people conveyed to me, directly and powerfully, their pain and suffering and their struggles with healing.

In going about my climate study, I did not conduct systematic interviews with particular groups of people as in previous research. Rather, I relied on the findings of scientists and knowledgeable observers, and on the recorded experience of survivors of climate events. But I brought to the climate work that longstanding exposure to raw human experience that I had gained through decades of interviews. What people went through in Hiroshima, Auschwitz, and Vietnam had become part of my enduring consciousness and a source of survivorlike feelings that have lasted a lifetime. To be sure, I have been a witnessing professional and not a survivor, but my interview method took me as close to survivor experience as a nonsurvivor can get. Climate change could never become an abstraction for me. Rather than a theoretical projection of a possible future, I came to see global warming as a source of ever-increasing human suffering.

3

Apocalyptic Twins: Nuclear and Climate Threats

Psychologically speaking, one of the terrible legacies of the twentieth century was our realization that we could annihilate ourselves as a species with our own technology. What resulted in our minds was *imagery of extinction*. The methodical quality of Nazi genocide contributed to that imagery, but nuclear weapons have been at its center. Imagery of extinction followed directly upon the atomic bombings of Hiroshima and Nagasaki, and the development of the hydrogen bomb rendered this vision potentially literal as we recognized our capacity to kill every last human being on the planet. As meaning-hungry creatures, we have struggled with ways to take in, to grasp and cope with, a world-ending catastrophe. Now global warming joins, encompasses, and interacts with nuclear weapons in the darkness of that apocalyptic category.

Albert Einstein once described the environment as

"everything that isn't me." He also saw nature as "a magnificent structure" that he and others were seeking to "comprehend," however "imperfectly," and which inspired "a feeling of humility" and "a genuinely religious feeling that has nothing to do with mysticism." One might add that even impressive human constructions—our automobiles and computers and airplanes, our buildings and bridges—originate in the materials of nature. Nature, then, is the earth itself, indeed the cosmos. That is why neither hydrogen bombs nor climate change can destroy nature. But they can threaten much life in nature, especially human life.

Living Out the Connection—Earle Reynolds

There are a few people who have lived out the connection between the apocalyptic twins, one of whom I had the good fortune to encounter in Hiroshima. Earle Reynolds was to become an iconic figure in his pioneering activist efforts to confront nuclear and climate threats. He was a physical anthropologist who was sent to Hiroshima as part of the Atomic Bomb Casualty Commission (ABCC), an American medical group that examined people there to determine the effects of the atomic bomb, notably those caused by radiation. Reynolds himself did a study on how these effects brought about impairment in the growth and development of children.

The ABCC was highly controversial in Hiroshima for at least two reasons: It was known to be an arm of the American Atomic Energy Commission, which also happened to be the agency responsible for nuclear weapons. And despite its close study of survivors it refused to offer treatment of any kind. There was concern (as expressed in an internal ABCC document) that such treatment would be understood as "an act of atonement for having used the weapons in the first place" that would "feed anti-American propaganda." As Susan Lindee, the historian of science who studied the ABCC, observed, "It was necessary, somehow, to redeem the bomb."

Reynolds came to believe that the ABCC was not only slow to report delayed radiation effects but in general understated or negated such effects in its public pronouncements. A detailed conversation I had in 1962 with the head of the ABCC in Hiroshima at the time convinced me that there was truth to Reynolds's suspicion. Allowing for scientific caution and genuine concern about causing anxiety in a sensitive population, I was struck by the commission head's tendency to minimize the impact of the bomb.

After leaving the ABCC, Reynolds decided to carry out a childhood dream of sailing around the world on a yacht. Hiroshima was the place where his yacht, the *Phoenix*, was built, and the impetus for the places Reynolds and his family would take it. In 1958, influenced by Quaker pacifists, the Reynolds family sailed into the nuclear testing area in the Marshall Islands, which had officially been designated as a forbidden zone for shipping. In 1961 the Reynolds family took the yacht into a Soviet testing area, and later, with a multinational crew, delivered medical aid to the Red Cross Society of North Vietnam for civilian victims of the Vietnam War.

I sought out Reynolds in Hiroshima in 1962 and met with him in a restaurant near the pier where his yacht was moored. As I wrote to my friend David Riesman, the noted antinuclear sociologist, I was strongly impressed by "how small" both the man (he was of average height) and the yacht (fifty feet long, as I later learned) were for a voyage into waters controlled by nuclear behemoths. There was nothing easy or smooth about Reynolds, whom I described as "determined, single-minded, inflexible, and mostly correct in what he had to say about the dangers of nuclear weapons and the need to do something about them." The voyages he and his family and crew took into nuclear test areas were a form of Quaker witness and protest.

Reynolds was an unusual combination of the most methodical and determined of men on the one hand, and the greatest of risk takers and adventurers on the other. Whether sailing into nuclear test zones, writing about the broader dangers of radiation, or creating a peace center in Hiroshima, he stubbornly persevered in seeing things through. Yet in his forays into danger and his surrender of personal and scientific privilege, he bore some resemblance to his parents and other kin who were circus performers, perhaps especially to his father and uncle who were noted trapeze artists and tightrope walkers. After being arrested for entering the forbidden waters of American nuclear testing, and tried and convicted in Hawaii, a commentator noted that "within two years Reynolds had gone from holding a significant position with a prestigious United States Government agency to being a convicted felon without a job and isolated from his family."

His marriage to Barbara Leonard Reynolds, herself a distinguished peace activist, ended a few years after their historic voyages, and he continued his journeys on the *Phoenix* with his second wife, Akie Nagami, a woman he met at Hiroshima Women's College while he was a visiting professor there. When they were forced by the Japanese government to leave Japan, they settled in the United States, where they worked in peace studies at the University of California, Santa Cruz, and other American universities, and in antinuclear activism.

I did not learn until much later that Earle Reynolds came to be seen as one of the direct forerunners of radical environmentalism on the oceans. He was particularly embraced by Greenpeace, a large and influential environmentalist group, as a pioneer for its "Mind Bombs," or dramatic confrontations on the high seas, whether on be-half of saving the whales or exposing and blocking oil rigs. The founding mission of the group was a 1971 voyage into a nuclear test zone in the Aleutian Islands off Alaska. Green-peace, the name given both the ship and the organization, suggests opposition to each of the apocalyptic twins. The Greenpeace strategy of Mind Bombs was said to have been influenced by Marshall McLuhan, the celebrated media theorist who emphasized the creation of lasting images that could significantly influence overall consciousness. Green-peace is still known for its nonviolent direct action on the high seas, which began with protesting nuclear testing and its poisoning of the environment. As Reynolds's story also makes clear, Hiroshima had much to do with those early antinuclear passions. This juxtaposition of nuclear and cli-mate threats can be found in most environmentalist groups, including the Sierra Club and Friends of the Earth.

The merging of the two threats is most dramatically epitomized by the concept of nuclear winter. Described by scientists in the 1980s, it suggests that a nuclear war could create debris from firestorms that would block the sun's rays and thereby render the global climate too cold to sus-tain human life. More recent work on nuclear winter varies in details but suggests that nuclear winter could occur in limited—that is, "marginal" or "regional"—nuclear war, and may require only fifty or so Hiroshima-sized bombs to bring it about. There has also been a new emphasis on "nuclear famine" caused by this climate disruption and other effects on agriculture, which could result in the deaths of as many as two billion people.

Keep in mind that nuclear winter is an ultimate form of devastation, going beyond that of the immediate explosions of nuclear war itself. And as its name suggests, nuclear winter is essentially a climate event, sometimes referred to as "global cooling." What has been insufficiently understood is the extent to which nuclear winter would bring the apocalyptic twins together to create a single, malignant, world-encompassing, climate-based annihilation of our habitat and its capacity to sustain human life.

The Marshall Islands—Lethal Convergence

The connection between nuclear and climate threats finds Job-like expression in one particular place: the Marshall Islands. Made up of twenty-four coral atolls comprising 1,156 islands and islets in the Pacific Ocean near the equator, the residents—now 68,480 people—have become victims of both nuclear and climate devastation. Told by the U.S. military governor (the islands were then part of a U.S. Trust Territory) that the nuclear testing was "for the good of mankind and to end all world wars," the Marshallese people were subjected to sixty-seven atmospheric nuclear bomb tests from 1946 to 1958. The most powerful of these was the Bravo test, detonated at Bikini Atoll in 1954, which unleashed the equivalent of one thousand Hiroshima bombs on the region in the largest detonation of any kind to have taken place on the face of the earth.

I was confronted with the vast dimensions of bodily and mental harm experienced by the islanders when asked, in 2002, to collaborate with my friend, the sociologist Kai Erikson, on a study of the social and psychological impact of the Bravo test on the people of the Utrik Atoll. The fieldwork was mainly done by Erikson, who traveled to the islands, but I worked closely with him on the report. We found that the islanders experienced almost certain increases in cancers and other physical illnesses (statistical proof is lacking because of the absence of systematic studies and the lack of medical facilities) and of birth defects (which have been documented). We reported that "the people of Utrik see themselves as so weakened, so damaged, so impaired by the effects of the bomb, that they are subject to all the troubles to which flesh is heir." And that "the land they live on, the food they eat, the water they drink and cook in, even the air they breathe is poisoned and unsafe for human life."

Their unending fear was that of invisible contamination, the same fear I observed among survivors of Hiroshima, a sense of being permanently vulnerable to a weapon that leaves behind in the bodies of those exposed to it deadly influences which may emerge at any time and strike down their victims. Like the people of Hiroshima, who described being treated like "guinea pigs," Utrik islanders spoke of American doctors who "came to study us," "came because they knew we were poisonous," and "used us like animals in an experiment."

The Marshall Islands lie only about five feet above sea level and, as Michael Gerrard, a leading authority on environmental law, explains: "As a result of sea level rise, the Marshall Islands and several other states face an existential threat"—that of their island home literally sinking into the sea—"an event unprecedented in the history of human civilization." Gerrard himself witnessed the literal merging of nuclear and climate threats during a visit to one of the atolls in 2013. He observed a three-hundred-foot concrete dome covering a thirty-three-foot crater into which had been poured extensive amounts of soil contaminated with plutonium, and plastic bags containing plutonium chunks, a structure so vulnerable that it "does not meet American standards for landfills for household trash." His inspection revealed that, because of sea rises and storms, the dome was deteriorating and was soon likely to be submerged or torn apart, "releasing its radioactive poison into the ocean and compounding the legacy our advanced civilization has left to this tiny island nation."

The islanders' living situation is in every sense dire: already subject to extreme droughts, they are unable to revert to traditional food crops because of poisons from the nuclear testing and a rise in sea level, and they suffer from acute shortages in drinking water. As the Marshall Islands foreign minister wrote: "For almost seventy years, my country, the Marshall Islands, has been fighting for its survival. . . . My people are not only thirsty and hungry, they are also getting sick. . . . [and] the situation is likely to get worse." Marshall Islanders are denied even the partial sense of rejuvenation experienced by Hiroshima residents in connection with the rebuilding of their city. (But Hiroshima too may be the victim of both climate change and nuclear weapons. In August 2014, it experienced a sequence of unprecedentedly extreme rainfall and multiple landslides, killing seventy people and destroying several districts of the city. A thoughtful commentator suggested that Hiroshima merits more attention as a potential center for studying "climate change and urban resilience.")

Now the Marshall Islands emerge as a collective example of witness and survivor wisdom: in 2011, Michael Gerrard helped organize an international conference covering all aspects of threatened island nations, which was the basis of his pioneering book on this subject. And, drawing upon the support of American sympathizers, Marshall Islands leaders mounted an international legal case in 2015 in which they took the initiative to sue all nuclear weapons–possessing nations for failing to pursue nuclear disarmament, as required for compliance with the Nuclear Nonproliferation Treaty of 1970. The case argued that the obligation to pursue such nuclear disarmament "in good faith" applied not only to the five signatories of the nonproliferation treaty— the United States, the United Kingdom, Russia, China, and France—but to the four who had never been or were no longer part of that treaty—Israel, Pakistan, India, and North Korea. Both the International Court of Justice and the U.S. Federal Court in San Francisco—the two places where the suit was filed—dismissed it on jurisdictional grounds, stating that they did not having the authority to try the case. But this Job of nations has made an important statement, which, according to the *Bulletin of the Atomic Scientists*, could have future influence on a "new politics of nuclear disarmament, a politics that challenges the very legitimacy and legality of nuclear weapons possession."

Nuclear Tests—Environmental Experiments

The exploding of nuclear bombs in what are called "tests" has always taken the form of a broader experiment. What has been mostly studied is the overall destructiveness of the weapons, with much attention given to their specific effects on the environment. For instance, there was a vast tree experiment taking place in conjunction with the 1953 nuclear tests in Nevada: imported trees were artificially anchored to create a "traumatized forest" in which tree and branch breakage and defoliation were closely monitored. In those and other tests, there was extensive study of the effects of radioactive fallout on human cells, plants, and animals, which enlisted the services of not only biologists but geologists and agriculturalists and others associated with the earth sciences.

From the start the testers were constantly concerned about how the weapons would affect the overall ecosystem, including the weather, and this concern led to the creation of a number of the institutions and concepts that have contributed greatly to scientific discovery about climate change. The irony is that all of the testing was done—insisted upon—for purposes of "national security," but the potential damage to the environment came to be seen as a threat to that same national security. The Limited Nuclear Test Ban Treaty of 1963 between the United States and the Soviet Union, which required that nuclear tests go underground to avoid harmful effects on oceans, land, air, and outer space, could be considered, as one observer put it, "an environmental protection treaty." Indeed, with nuclear testing, "the world itself became the laboratory."

The most chilling rendition of nuclear world-ending is portrayed in *On the Beach*, a 1957 novel by Nevil Shute, and in a devastating 1959 motion picture based on it. In that narrative, a nuclear war has destroyed the northern hemisphere, and the released radiation makes its way to southern Australia, gradually inundating all of the earth's atmosphere and resulting in the death of all human beings, although most of the film's main characters take their own lives in order to exert some control over how they die. In the novel and film and discussions about them, radiation is constantly referred to as "pollution" and experienced as an environmental poisoning. From the beginning, that is, nuclear threat has been inseparable from environmental issues, and later projections of global warming have reaffirmed this tandem scenario of species danger.

No wonder that militarized scientific minds projected ways of combining nuclear and environmental assaults in what Jacob Darwin Hamblin called "Arming Mother Nature." Following World War II, scientists participated in the Cold War between the United States and the Soviet Union by working on radiological contamination, biological weapons, weather control, and "several other projects that united scientific knowledge of the natural environment with the strategic goal of killing large numbers of people." It was as if the atomic bombings of Hiroshima and Nagasaki crossed a threshold into limitless applications of the science of destruction. This "catastrophic environmentalism" included projections during the Korean War of spraying plutonium waste across Korea to create a "death belt," using hydrogen bombs to trigger earthquakes, spraying yellow fever across Soviet cities, and melting the arctic ice caps with millions of tons of soot from atomic explosions. At first I wondered whether this was more a matter of destroying than "arming" Mother Nature, but I came to realize that Hamblin was correct in his description because these projections involved the conversion of nature into deadly weaponry. The extremities of World War II extended into a Cold War mentality that deepened the quantum leap into the science and technology of killing.

The Merging of World-Ending Threats

It is difficult to grasp the impact of these draconian projections, many of them confined to limited scientific and military circles. But I learned much about the responses of ordinary people in research I did with my close colleague and friend Charles Strozier in the early 1990s. Our extensive interview study, supported by the MacArthur Foundation, involved different groups in American society and was entitled "Nuclear Fear and the American Self." The frequency of environmental and climate anxieties took us by surprise. So did the close interweaving of climate anxieties with nuclear anxieties. We now believe that these findings reflected a shift in the content of the imagery of extinction, from nuclear to climate fear, and to a psychological representation of such world-ending imagery in which distinctions between the two could be lost.

For instance, a young, unemployed African American man spoke first of the total devastation depicted in the film about nuclear war called *The Day After*, and immediately associated that with extreme environmental pollution: "A lot of things is gettin' extinct, zoos and stuff like that. . . . The world is dying, looks like, ya know? . . . Big cities get pollution, do they call it greenhouse effect? . . . New York City, Detroit, all those large cities and stuff like that. It causes acid rain. . . . All these chemicals is gettin' locked in the sky and rain mixes and building up all kinds of shit and then coming back down, skin cancer and shit like that."

Another man similarly began talking about nuclear war: "Roaches will still be alive. . . . We're the ones who did it." He then slipped seamlessly into a discussion of the destruction of our habitat: "We're the ones who contaminated everything that was good. Ya know, like what they're doing to the ozone layer. Winter now is like summer, summer is like winter. . . . Why is it so hot? Why is this happening?" And he added: "It's over! It's over!"

These responses demonstrate how specific threats, in this case nuclear and climate threats, can be subsumed into an overall psychological experience of world ending. Also blended here were the extreme pollutions of the time, the ozone layer and acid rain. All of these psychological associations could occur within the same thought, sentence, or phrase.

The Corrupt Scientific Connection

I encountered another kind of continuity between nuclear and climate threats that surprised me. I found that a number of scientists and strategists who embraced nuclear weapons went on later to falsify global warming. Striking examples here are Edward Teller and Herman Kahn. Teller was a brilliant physicist who had much to do with the creation of the hydrogen bomb and came to exemplify the spiritual disease I call *nuclearism*. Nuclearism involves an exaggerated embrace of the weapons, sometimes to the point of deification, equating them not only with national security, but with keeping the peace and even keeping the world going. Teller combined belief in the unlimited capacity of technoscience with a totalistic form of anticommunism, insisting that "if we stop [building and stockpiling ever more destructive and accurate nuclear weapons], we are falling behind. . . . We cannot and must not try to limit the use of the weapons."

But it was this accomplished scientist's comments on radiation in 1975 that were astonishing: "Radiation from test fallout might be slightly harmful to humans. It might be slightly beneficial. It might have no effect at all." Equally astonishing were his strangely similar comments on climate change twenty-three years later: "Society's emissions of carbon dioxide may or may not turn out to have something significant to do with global warming—the jury is still out." Teller opposed international agreements designed to reduce climate change, and then became an outspoken advocate of a grandiose form of geoengineering that would solve the climate problem by introducing into the stratosphere enormous amounts of fine particles of a kind emitted by volcanoes, in order to cool the effects of the sun. Most scientists view this proposal as impossible to carry out and potentiality dangerous, even if some concede that lesser forms of geoengineering might be worth investigating.

Herman Kahn was trained as a physicist but functioned primarily as a nuclear strategist. Also an ardent nuclearist, he envisioned fighting, surviving, and winning nuclear wars, even if they involved (he added unironically) "somewhat larger risks than those to which we subject our industrial workers in peacetime." To help overcome fear of radiation and the psychological contagion of fear, Kahn recommended that each person have a meter for registering the radiation level, with the bizarre psychological rationale: "Assume now that a man gets sick from causes other than radiation. Not believing this, his morale begins to drop. You look at his meter and say, 'You've only received ten roentgens. Why are you vomiting? Pull yourself together and get to work.'"

Kahn is viewed as the father of scenario planning, the projections of give-and-take behavior by nuclear protagonists. By viewing nuclear war as not only thinkable but winnable, Kahn did much to contribute to the nuclear normality I will discuss in chapter 5.

Kahn was a futurist, and in his last years became involved in environmental issues. He co-edited the book *The Resourceful Earth*, meant to counter the Global 2000 Report to the President, a careful three-year study completed in 1980 that predicted "If present trends continue, the world in 2000 will be more crowded, more polluted, less stable ecologically and more vulnerable to disruption than the world we live in now," and that "serious stresses involving population, resources and environment are clearly visible ahead." Kahn and his co-editor reversed that prediction, declaring that "if present trends continue, the world in 2000 will be less crowded (though more populated), less polluted, more stable ecologically, and less vulnerable to resource-supply disruption than the world we live in now." They insisted that global warming was a "normal oscillation" of nature and the dangers were greatly exaggerated, and that all could be managed by technological and scientific advance.

Teller and Kahn combined an absolute faith in science and technology—in what I call *rescue technologies*—with passionate anticommunism and advocacy for unlimited American power, thereby contributing to the most dangerous expressions of nuclearism. They could then cease to acknowledge or feel—and instead reject, deny, and falsify—the world-destroying effects of both nuclear weapons and global warming.

Imagining the Real

Even scientists who would later renounce nuclearism went through periods of *psychic numbing*, a diminished capacity or inclination to feel. Those who worked on the bomb at Los Alamos, New Mexico, were described by a knowledgeable observer as having undergone a "half-conscious closing of the mind." The truly remarkable physicist Richard Feynman put it well in reporting his own and fellow physicists' experience:

> You see what happened to me—what happened to the rest of us, is we started for a good reason [beating the Nazis to the creation of the atomic bomb], then you're working very hard to accomplish something, and it's a pleasure, it's excitement. And you just stop thinking, you know, you stop.

Feynman rightly stresses the fascination of the intellectual task, along with the consuming motivation of winning the nuclear race with Nazi Germany. They, like Teller and Kahn later, could not afford to take in, to *experience*, what a nuclear weapon would do to thousands, or hundreds of thousands, or millions of human beings. There are parallel patterns of psychic numbing in those who falsify or reject, or refuse to act upon, unchecked climate change.

Everyone calls forth a measure of psychic numbing toward nuclear and climate threats. Such numbing, as I observed in earlier work on the effects of nuclear weapons, has to do with the mind's resistance to the unmanageable extremity of the catastrophe, to the infinite reaches of death and pain. The numbing also has to do with the lack of any prior image or model that could help us grasp the nature of the threat. In Hiroshima, for instance, people who survived the blast were at first deeply confused about what had happened to them and to their city. They wondered whether it was a huge "electric short," a form of "Buddhist hell," or "the end of the world."

We have similar difficulties imagining the effects of advanced global warming. We find ourselves continuously searching for—and also resisting—images that might help us to take in what is also an unprecedented phenomenon. Even when exposed to melting glaciers, severe floods, droughts, and wildfires, it is hard for us to find in them clear models of what we can expect.

With both nuclear and climate threats, that is, there is a near impossibility of, in Martin Buber's phrase, "imagining the real." No wonder we tend to see each of them as beyond description or comprehension, as driven by otherworldly forces that render us tiny and helpless, rather than as lethal mechanisms we ourselves have created and are quite capable of understanding. Even if nuclear and climate threats had no other similarities, their insistent calling forth of unmanageable imagery of extinction would in itself render them of a piece. Here one must recognize that psychic numbing in general is a human equivalent to the way animals "freeze" (sometimes called "playing dead") when threatened and lacking a path of either resistance or escape. The widespread numbing created by nuclear and climate imagery of extinction can be understood as playing dead on the part of the majority of people on earth.

Narratives of "the End"

In that struggle to "imagine the real," many turn to ideas and narratives associated with religion, especially apocalyptic religion. In most Jewish and Christian apocalyptic narratives there is a sequence of human misbehavior and violation or neglect of God's commandments, and God's furious retribution in some sort of world destruction in which only a small "remnant" of believers survives, leading to a new age of spiritual rebirth with God once more in charge of his followers—in the Christian version through the return of Jesus.

Nuclear holocaust lends itself directly to the apocalyptic narrative of the Book of Revelation, especially its depiction of the "lake of fire" that is "burning with brimstone," where sinners go to a dreaded "second death." Inevitably, believers have inserted nuclear weapons into that sequence, as the vehicle that sets it in motion. In that way weapons could be seen as contributing to the cleansing and purification promised by the apocalyptic narrative. But the weapons and their radiation effects can also be understood to be expressions of human sinfulness that must themselves be eliminated by the apocalypse. Some have wondered whether God would want to impose such suffering, or whether he would wish to leave the earth in such terrible shape for Jesus's return. Fundamentalists may hold contradictory views on these matters, while keeping to a belief in the basic connection between nuclear weapons and the apocalypse. Hence the tendency of certain fundamentalists to welcome nuclear holocaust because of the apocalyptic promise of spiritual rebirth, so that "the earth shall become heaven."

In my work in the late 1990s on Aum Shinrikyo, the

fanatical Japanese cult that released sarin gas in Tokyo subway stations, I encountered a more lethal impulse to actively participate in apocalyptic destruction, to engage in violence that might hasten the desired spiritual purification. Ancient Hebrew prophets called this "forcing the end," and tended to reject it in favor of permitting God to control the timetable. Aum's guru, Shōkō Asahara, showed no such restraint. He had his cult actually produce crude versions of chemical and biological weapons while seeking (fortunately, unsuccessfully) to acquire nuclear weapons. Aum brought forth the combination of ultimate fanaticism and ultimate weapons.

It is significant that Aum Shinrikyo, founded on early Buddhist, and to some extent Hindu, ideas, nonetheless seized upon the Christian apocalyptic narrative. The Islamic State (or ISIS) also envisions an apocalyptic end of the world, in its case through the creation of a caliphate with sharia law as necessary preparation for that apocalypse. ISIS has been primarily an army, but embraces an extreme Islamic apocalyptic vision, while Aum Shinrikyo has been primarily a religious cult that embraced a murderous military path, each in a quest for the death and rebirth of the human species.

I have argued that ancient apocalyptic narratives, along with political versions of them, become more attractive in a nuclear weapons–centered world. Once it is known that we can destroy our species with our own technology, there is a profound temptation to provide such an ending with a meaning structure, and what better meaning structure than an apocalyptic narrative? In that sense Aum and ISIS, along with many other groups, express not just a biblical but a contemporary version of nuclear-propelled end-time inclination and imagery, with the ever-present possibility of a proactive plunge into "forcing the end" by joining in mass violence.

Present-day leaders of nuclear weapons–possessing countries may not be completely immune from such inclinations, whether their arsenals are American, British, Chinese, French, Indian, Israeli, North Korean, Pakistani, or Russian. Without being generally fanatical, such leaders could be drawn to nuclearism and a version of the apocalyptic story, which is all the more reason to strengthen whatever technical and ideological restraints are already in place.

In all this we must recognize the phenomenon of the "smaller apocalypse"—the experience of the relatively local disaster as apocalyptic. I observed that pattern in connection with my work with survivors of the Buffalo Creek, West Virginia, flood disaster of 1972. The flood was created by corporate negligence: the dumping of coal waste in a mountain stream, resulting in the bursting of an artificial dam that caused the death of 125 people and rendered about five thousand more homeless. Although a relatively limited event, it encompassed an entire area, or "hollow." And a number of people expressed world-ending experiences such as: "It was the end of time" and "Everything came to an end, just stopped. Everything was wiped out."

Similarly, people trapped in the Twin Towers during the terrorist attack on New York City in September 2001 invoked an "atomic, mushroom cloud," and spoke of Hiroshima. Those associations to nuclear weapons and the end of the world could occur in many different disasters that signified ultimate destruction and the need for remaking of one's own particular shattered world. Given the influence of climate change in creating more regional disasters, we may speak of apocalypse in a thousand cuts—a dynamic of world-ending visions intensifying one another and often including Hiroshima as an expression of infinite human-caused destruction so vast as to seem mysterious and biblical.

In one sense, climate change, with its slower, incremental sequence, lends itself less to the apocalyptic drama. But when one looks more closely at biblical narratives, they are infused with climate catastrophe. The story of Noah and his ark is perhaps the best-known climate narrative, and came about because of God's declaration in the Book of Genesis that "I will cause it to rain on the earth forty days and forty nights." In addition to floods, God's retribution in various parts of the Bible takes the form of hailstorms, droughts, "darkness over the land," and lethal earthquakes. In the Book of Revelation, the most apocalyptic New Testament text, there is a great earthquake in which "the sun becomes black as sackcloth of hair and the moon like blood, mountains and islands are moved out of place, the rivers run red with blood, and the earth itself is scorched." We can say that the Bible begins (Genesis) and ends (Revelation) with assaults on the human habitat. That is the way God destroys all sinners, along with their sins. Biblically speaking, the human habitat is the essential apocalyptic target. And contemporary Christian writers in the apocalyptic mode have readily made the connection between these biblical "extreme events" and present-day climate change. Climate change furthers the temptation to replace history with eschatology.

Which raises the question of the pervasive attraction of the apocalyptic narrative. We have seen that it can be called forth not only in response to nuclear war and climate catastrophe but also to relatively limited disasters. It can in fact be operative for many even in the absence of extreme events, as an apocalyptic vision is never too far removed from a Christian worldview. Most basically, the narrative is a response to our knowledge that we die, infusing that knowledge with meaning. It promises a profound sharing of death and forms of redemption that, at least in the Christian version, liberates one from the prison of sin. Most important, it holds out the promise of spiritual renewal. Since the apocalypse always includes a remnant that remains alive to regenerate and repopulate the newly purified world, it becomes one of the great survivor narratives. Whether or not one imagines oneself part of this remnant, one's death takes on special value, even sacred meaning, as a contribution to God's creation of a human future. In that way the apocalyptic narrative takes its place within the pervasive mythological theme of death and rebirth.

Both nuclear and climate threat, then, are readily joined with biblical apocalypse. The climate version, though slower to unfold, finds more biblical models. But there are important distinctions to be made in ways the mind responds to each of the two threats.

4

Different Mental Struggles: Nuclear and Climate Truths

Nuclear and climate threats are also separate and different from each other. With nuclear weapons, the mind must contemplate specific *things*—bombs that bring about revolutionary dimensions of blast, heat, and radiation. The imagined catastrophe is immediate and decisive, and the atomic bombing of Hiroshima and Nagasaki provides a model of unprecedented slaughter and suffering. With climate threat, the mind encounters no new physical entities or things, but rather an incremental sequence of an increasingly inhospitable habitat, a progression rather than an explosion, and a series of projections of what can be expected, as opposed to a clear display of ultimate destruction.

Ecological Dissonance

I have suggested that the claim that we humans lack the hard wiring, the evolutionary brain capacity, for environmental projections writes us off too quickly, and that imagining the future is a key to the extraordinary human capacity for adaptation. That imagined future can include a vision of ourselves as survivors of disasters to come and finding survivor meaning in struggling to prevent those impending disasters. Any such adaptation is neither static nor completely predictable. The mind's returns are far from in, and here one must again recognize climate change as the most daunting task ever faced by humankind.

There is much talk about the cognitive dissonance that prevails between the still quite functional environment (for most of us) and the forthcoming lethal environment projected by climate scientists. One could go further and speak of *ecological dissonance* in our relationship to the natural world: the contrast between the classical sense of being nurtured by Mother Nature and present awareness of dangerous interaction with our natural habitat. Any such dissonance can lead to a rejection of unpalatable climate truths.

These mental struggles can cause us to feel deeply insecure in our natural home and experience ourselves as floating, as lacking grounding for ourselves and our cultural institutions. We take on some of the Hiroshima survivors' early fear that nature would dry up altogether in their city. In Hiroshima that fear was in some measure assuaged by the reappearance of cherry blossoms the following spring, and by early efforts at rebuilding. In contrast, we experience climate change as inexorable, with continuous decimation of what we humans need from nature. We may speak of a creeping poisoning of atmosphere and oceans, with clear evidence of lethal human contribution.

"Humane" Nuclearism?

Scientists have played a crucial role in both nuclear and climate issues, but have differed dramatically in their relationship to the two threats. Climate scientists did not, like nuclear scientists, create the instruments that could destroy humankind, and did not take on the very complex legacy of their counterparts. As we have seen, that legacy included feelings of responsibility and guilt for having brought species-threatening weapons into the world and also, in the case of many, for having wanted them to be used. They had been part of a moral crusade to produce the weapon before it could be constructed by Nazi scientists, but the crusade had continued after it was officially learned that the Nazi bomb project had failed. Also contributing to those feelings of responsibility and guilt were the nuclear scientists' deep satisfaction, even transcendent pleasure, in their selfless collaboration in successfully achieving a very difficult common goal.

The brilliant physicist Hans Bethe described to me the "very cooperative" atmosphere at Los Alamos, the intense camaraderie under the intellectually brilliant and personally benevolent leadership of the highly admired Robert Oppenheimer. The Los Alamos experience was a heroic moment for many of those physicists, who (as a prominent colleague put it) later engaged in "endless talk about [those] days . . . through [which] shone a glow of pride and nostalgia . . . [about] a great experience, a time of hard work and comradeship and deep happiness." Nuclear scientists became part of the ultimate irony of experiencing a utopian community while creating the cruelest weapon in human history.

One could say that the scientists underwent a form of doubling: while aware of participating in a moral crusade to win the war against an evil enemy, they also understood in at least a part of their minds that they were creating something infinitely destructive and dangerous to humankind. As early as 1942 some of them wondered whether the bomb would ignite sufficient nitrogen in the atmosphere or hydrogen in the oceans to burn up the planet, and in July 1945 just before the Trinity test, Enrico Fermi offered to take bets from colleagues on whether the bomb "would merely destroy New Mexico or destroy the world." There was a bit of gallows humor there, but at the same time General Leslie Groves, the military head of the project, ordered the preparation of a just-in-case public statement explaining the sudden death of a large number of scientists, and informed the governor of New Mexico that he might have to declare martial law.

Inevitably, those working on the bomb were consumed with scientific curiosity about their product, by pride in the success of their creation—a device that exploded, as opposed to a dud or a failure. Many succumbed to the fallacy of what one might call "humane" nuclearism: advocacy of use of the bomb as a weapon in the war with the expectation that its revolutionary impact would convince world leaders that they must pursue peace. Leo Szilard, for instance, was to become one of the most outspoken antinuclear voices among them, but he did much to initiate the project in early 1944 and wrote: "It will hardly be possible to get political action [for lasting peace] unless high efficiency atomic bombs have actually been used in this war and the fact of their destructive power has deeply penetrated the mind of the public."

Robert Oppenheimer, who would later oppose the development of large hydrogen bombs, was a consistent advocate for the use of the atomic bomb. During the weapon's creation at Los Alamos he was strongly influenced by Niels Bohr, a much-respected older-generation Danish physicist.

The two men explored the idea of applying Bohr's theory of complementarity to the use of the bomb. In physics (quantum mechanics) the theory holds that objects have complementary properties that cannot be measured accurately at the same time, a vivid example being those of a particle and a wave. Similarly, Bohr and Oppenheimer reflected, the bomb itself could bring about great destruction but also human redemption in the form of future international cooperation and the prevention of war.

What one can also say about these remarkable reflections is that, under certain circumstances—in this case scientists seeking to devise a weapon that would win a war against a murderous Nazi regime—the most brilliant and humane minds can be drawn to nuclear weapons as an antidote to other forms of violence. In the process, what is perceived as a moral imperative can engender powerful psychic numbing and self-deception. To be sure, such advocacy could be replaced, quickly or slowly, by articulate opposition to the weapons. But the dangerous lure of various forms of "humane" nuclearism, of finding potential salvation in the very power of nuclear weapons, remains with us.

Prophetic Survivors

A few of the scientists involved in the Manhattan Project, Leo Szilard at Los Alamos and James Franck in Chicago, tried desperately to organize their colleagues to prevent the use of the weapon on a human population. And Eugene Rabinowitch told of walking through the streets of Chicago, together with a few other scientists working with him on the project, and "vividly imagining the sky suddenly lit up by a giant fireball, the steel skeletons of skyscrapers bending into grotesque shapes . . . until a great cloud of dust rose and settled into the crumbling city." This could be understood as an anticipated or prospective survivor-like experience. Survivors seek a measure of meaning in what they have been through. For Rabinowitch, that had to do with truths about the weapon and possibly a sense of self-condemnation for having created it. Rabinowitch immediately intensified his work on the Franck Report, which opposed the use of the weapon. Then in July 1945 the scientists witnessed the Alamogordo test and many of them underwent a kind of conversion in the desert as they experienced the bomb's near-cosmic dimensions and committed themselves to warning the world about the dangers of what they had created. Hans Bethe described to me how photographs of Hiroshima, shown to scientists a few days after the bombing, made "an awful impression," and how those scientists "were really horrified."

A few months later, when Oppenheimer made his famous statement that "the physicists have known sin," he was referring to what he called their "peculiarly intimate responsibility" in creating atomic weapons whose use "dramatized so mercilessly the inhumanity and evil of modern war." He was not saying that dropping the bomb was wrong. Rather, he was saying that physicists could not escape engaging in evil, however justified their project. But I would add that the evil—or the sense of it—concerned not only the making of the weapon but undergoing deep pleasure and satisfaction in the process.

These complex feelings helped fuel the physicists' early antinuclear movement. They experienced an animating relationship to guilt, their self-condemnation transformed into a passionate sense of larger responsibility to the world. I have referred to these scientists as *prophetic survivors* of the catastrophic destructiveness of their own creation. Their many-sided connection to that creation gave a special character, an invaluable authority and edgy intensity, to their witness and to the movement they created.

Climate Change—the Scientific Message

Climate scientists differ from their nuclear counterparts in having done nothing as a group to create the problem. But like nuclear scientists they have done everything to identify the source of the danger, including the crucial human contribution to climate change through the promiscuous use of fossil fuels. Even more than nuclear scientists, they have been virtually alone in their struggle to make known the dire significance of their findings and to convince others of the extraordinary danger those findings convey and of the necessity to act.

Climate findings, moreover, are no longer just a matter of computer projection—what *would* happen in the future as a result of our spewing carbon and methane into the atmosphere and the oceans. During recent years, climate scientists have been recording readings of dangerous global warming that has already occurred, of the increasingly hot years on our planet and the melting of huge glaciers in the arctic, and of widespread droughts, wildfires, and flooding. Surely climate scientists have also been prophetic survivors, responsible for virtually all knowledge about global warming and the kind of threat it poses to the human future.

But their scientific message has probably been slower to reach the general public: this is partly because most of the more malignant climate effects are not yet widely evident; partly because corrupt scientists and their equally corrupt financial sponsors have denied, rejected, and falsified climate truths in ways that have sowed widespread confusion; and partly because researchers have observed traditional scientific caution by understating, scrupulously reporting uncertainties, and emphasizing the difficulties of prediction.

It has also been pointed out that few American climate

scientists have had the highly cosmopolitan social perspectives of the earlier generation of physicists associated with the atomic bomb, including as it did Niels Bohr, Leo Szilard, Joseph Rotblat, Robert Oppenheimer, Eugene Rabinowitch, Philip Morrison, and James Franck. But there have emerged new generations of climate scientists who combine professional observations with outspoken and articulate advocacy of policies to curb global warming.

Here I would mention two key examples, James Hansen and Naomi Oreskes. Hansen, a physicist and astronomer who for many years headed the NASA Goddard Institute for Space Studies, gave testimony to the U.S. Senate Committee on Energy and Natural Resources, in 1988, that profoundly influenced public awareness of global warming. Hansen called forth extensive data to support three powerful conclusions: global warming was already occurring, as "the earth is warmer in 1988 than at any time in the history of instrumental measurements"; the warming is due to the "greenhouse effect," or the trapping of solar radiation caused by the presence in the atmosphere of carbon dioxide and methane, which absorb heat from incoming sunlight; and this greenhouse effect has increased the probability of "extreme events"—summer heat waves and droughts—and will do so at a greater rate in subsequent years. Hansen's testimony was a turning point in what would become the climate swerve, and opened the way to establishing the human contribution to dangerous warming. Hansen also created a model of a scholar-activist willing not only to speak boldly about findings and advocacies, but to be arrested for participation in nonviolent resistance to climate policies he considered dangerous.

Naomi Oreskes is both a geologist and historian of science, whose book, *Merchants of Doubt*, written with Eric Conway, exposed the phenomenon of climate denial as a carefully constructed product of fossil fuel interests. The book showed how those interests followed an earlier model created by tobacco companies to counter the clear scientific evidence of the harmful effects of tobacco on health. The model included supporting spurious institutes and corrupt scientists, some of whom went directly from denying the harmfulness of tobacco to climate denial, and who constantly fed skeptical reports to the general public. American media could collude by insisting on presenting "both sides" of the question, despite the overwhelming scientific consensus on global warming, as documented particularly by Oreskes in her tireless writing and speaking.

If climate scientists mostly speak in a softer voice than their nuclear colleagues, it is largely because of the incremental nature of their subject and the greater room for divergent findings, not about the basic truth of global warming but about when and in what form its dangerous effects can occur. What those same climate scientists have achieved is quite remarkable. It is they who brought to the world the urgent message about human-caused threat to our human habitat. And it was their computer projections about levels of carbon in the atmosphere that directly resulted in a major climate activism movement. The name of that movement, 350.org, derives from the parts per million of carbon dioxide in the atmosphere considered a relatively safe upper limit by James Hansen on the basis of research at the Goddard Institute.

"Humane" Carbonism and Climate Nuclearism

Can we identify a form of "humane" carbonism as a parallel to "humane" nuclearism? Perhaps we should not be surprised at declarations that "Carbon dioxide is green!" by climate falsifiers who seize upon the temporary benefits it can sometimes offer to plants, and view themselves as "debunkers" of "climate change myths." More puzzling is the position of physicist and public intellectual Freeman Dyson, one of the great scientific minds of his generation and an ardent advocate of nuclear disarmament. Dyson expresses the hope that "the scientists and politicians who have been blindly demonizing carbon dioxide for thirty-seven years will one day open their eyes and look at the evidence." In 2015 he wrote an introduction to a book entitled "Carbon Dioxide: The Good News," affirming the title and insisting that "the non-climatic effects of carbon dioxide as a sustainer of wildlife and crop plants are enormously beneficial . . . [and] the possibly harmful climatic effects of carbon dioxide have been greatly exaggerated, and . . . the benefits clearly outweigh the possible damage."

I have no certainty about what motivates Dyson to so flagrantly violate principles of scientific evidence, as he has over decades on this issue. One possible factor is the impulse of the brilliant contrarian, who delights in upending elements of conventional wisdom in his field and claims the right to see everything through the prism of his own original mind. Also relevant may be the sense of nature as life's ultimate source, so ingrained in all of us, perhaps especially strong in a physicist like Dyson who may resemble Einstein in his awe of nature. And there may as well be the seemingly opposite sense of science prevailing over "nature" in a large-scale rescue technology, consistent with a statement he once made to the effect that we could pretty much fix things by creating more snow in the arctic.

Another dangerous tendency is what can be called *climate nuclearism*, the advocacy of widespread use of nuclear power as a "clean" alternative to fossil fuels. The leading scientific voice for climate nuclearism has been James Hansen, the same man who has so admirably investigated climate dangers and made these truths known to the American people. Hansen has expressed his nuclear advocacy with considerable passion, but in ways that critics, including fellow scientists, believe overstates nuclear safety and understates the dimensions of nuclear accidents that have already occurred.

I stress the simple but telling point that nuclear energy and nuclear weapons make use of the same radiation-centered technology, whatever the efforts to view them as separate. That is why the use of nuclear power plants tends to facilitate the creation of nuclear weapons. Both entities, moreover, produce radioactive waste that can pose extensive danger for millions of years, and for which scientists and engineers have found no satisfactory means of safe disposal. Indeed, disposal of nuclear waste has been viewed as the Achilles' heel of the industry. More generally, I believe that radiation-centered nuclear technology is the most dangerous of all our technologies. Despite claims that technoscience can improve the safety of reactors, the possibility of human error can never be eliminated. Nor can the potentially catastrophic influence of extreme weather, as was the case with the Fukushima Daiichi nuclear disaster of 2011 in Japan. And even if risk assessment studies "prove" the rarity of nuclear accidents, those accidents can be catastrophic both in their immediate effects and their danger to future generations.

Two such examples are the meltdowns at Fukushima and at Chernobyl in the Ukrainian area of the then Soviet

Union in 1986. The Fukushima disaster, precipitated by an earthquake and a tsunami, resulted in what has been estimated to be 1,656 "indirect deaths" associated with the evacuation of an estimated 100,000 to 160,000 people because of high radiation levels in the area. The disaster threw corporate and government authorities into their own tsunamis of misinformation and cover-up. It spilled more toxic nuclear waste into the ocean than any other nuclear event, and is still doing so six years later.

Chernobyl, the worst nuclear power plant disaster in history, killed 28 emergency workers through acute radiation effects, and the number of cancer deaths related to the event worldwide was estimated by the Union of Concerned Scientists at approximately 27,000, and by Greenpeace at as many as 200,000. A nineteen-mile area surrounding the plant has been designated as a "zone of alienation," and has remained virtually uninhabited, with radiation levels so high that it is estimated to be unsafe for human life for another twenty thousand years.

Most of the studies of these two catastrophic events say relatively little about the terrible psychological and social damage, including the destruction of communities and the fear of radiation effects that could extend to people hundreds or even thousands of miles away. For those living nearby there was a sense that their own bodies had become part of the poisoned landscape. Pictures of what remains at

The Climate Swerve

Chernobyl still evoke a powerful sense of desolation, and those of Fukushima a sense of unending radiation disaster. For many people throughout the world, nuclear power is associated with danger, suffering, and desolation.

I was able to get a more intimate sense of a nuclear power disaster from work that Kai Erikson and I did at Three Mile Island in Pennsylvania just months after the accident, making use of an interview protocol that he and I had devised. Three Mile Island was the largest American nuclear power plant disaster, with severe damage to a reactor core. There were no immediate deaths and investigators have not identified a clear statistical increase in cancers or other bodily effects in the area, though some believe that such an increase might be revealed by research methods more sensitive to the effects of low-level radiation.

What cannot be disputed was the extreme anxiety in those exposed concerning possible effects of invisible contamination. People we interviewed spoke of their fear of developing cancer or leukemia and a number of them thought of Hiroshima and Nagasaki: "That's the same thing as what happened over in Japan . . . years later people died from it." And as one mother put it: "It's very upsetting to know that we may . . . become a statistic in a journal or a newspaper article twenty years down the road. . . . It can contaminate my kids . . . contaminate anybody that they decide to marry. . . . Are they going to have children that are disfigured or become diseased?" I was struck by the extent to which they echoed Hiroshima survivors in saying things like: "It was just that you . . . couldn't feel the radiation, you couldn't see it, you couldn't smell it or anything. . . . So you don't know if there's any radiation out there."

I spoke of this reaction of people at Three Mile Island as one of inner terror, a sense that one's body has been penetrated by a poison that could produce lethal forms of cancer and grotesque deformities in subsequent generations in an endless chain of life-threatening bodily damage. Many still considered their environment to be dangerous and were suspicious about the effectiveness of the official cleanup period. They also experienced anxiety and other post-traumatic symptoms in connection with reminders of the disaster. As one man put it: "It's like a monster in our backyard. Every day I can walk out of my house and look across the ball field and see these towers [the nuclear reactors]. That's a constant reminder . . ." Overall they were describing what is called Radiation Response Syndrome, likely to occur among people exposed to nuclear weapons or reactor meltdowns.

Above all people felt that authorities were contradictory, often untruthful, and confused. Concerning whether or not to evacuate, as one woman put it: "Reports we heard were so conflicting that we began to realize that there was no way we were going to find out what the truth was, or what was going on. Government officials saying 'Get out!' and on the other hand, owners of the plant saying 'Stay! There is nothing wrong. We've got it in control.'" I wondered whether nuclear accidents, once they occurred, could ever be "in control."

Yet as late as 2014 the Nuclear Energy Institute put out a newsletter for the industry that cheerfully reported on the positive "lessons" from the Three Mile Island accident, claiming that "the industry and the government responded swiftly and decisively." The newsletter pointed out that there were no injuries or deaths and "no measurable health effects," surely a dubious lesson. And it emphasized the industry's focus on the "highest levels of safety and reliability in the operation of nuclear power plants." The implication is that all is fine with the 104 operating reactors in this country and that nuclear accidents are a matter of the past—although both Chernobyl and Fukushima occurred between the 1979 Three Mile Island accident and the 2014 report.

Indeed, through the year 2014 there have been fifty-six significant nuclear accidents in the United States, with deaths recorded in at least seven of them. One of the non-lethal events was an emergency shutdown at the Pilgrim Plant in Plymouth, Massachusetts, serving the Cape Cod area where I have lived for a portion of each year since the 1960s. The plant has had many problems, including leaking steam valves that required a shutdown as recently as December 2016, and those of us in the area have long been aware of the impossibility of effective evacuation along the narrow Cape Cod peninsula in the case of a major accident. We have been heartened by the decision to close the reactor in 2019 but remain concerned that damage could be done before that.

Globally, there have been more than one hundred serious nuclear accidents as of 2014, fifty-seven since the Chernobyl disaster, and 60 percent of all of them in the United States.

The day after the Three Mile Island accident, I happened to be in Paris doing research on Nazi doctors and Auschwitz survivors. I attended a reception where I encountered French professionals and mid-level government bureaucrats, all of whom were upset by the nuclear accident. What they said was: "Those stupid Americans! They had to ruin everything. That could not happen in France." They were referring to the large French commitment to nuclear power, then envisioned to become the major source of energy for the country, and the French pride in its ostensibly accident-proof technology. But by 2001 there had been at least twelve significant nuclear accidents in France. And in 2016 it was discovered that the large steel forgings used in French plants had carbon-content irregularities that weakened the steel, and that there had been, according to the *Wall Street Journal*, "a decades long cover-up of man-ufacturing problems." Many plants had to be shut down for safety checks, and by October 2016, twenty of France's fifty-eight reactors were offline. There were also perennial nuclear waste–disposal problems, along with strong reactions by French citizens to Chernobyl and Fukushima, resulting in an active antinuclear movement, directed at an industry said to be in deep economic trouble.

The movement for nuclear power did not begin with environmental concerns, but rather with a vision of "atoms for peace," the term President Eisenhower used in a speech he gave in December 1953. This early example of a claim to "humane" nuclearism applied both to the new industry of nuclear power and to the idea that nuclear weapons would be a means of keeping the peace. The nuclear power industry then was launched as a kind of savior that would provide cheap and limitless worldwide energy supplies, and also as a redeemer of nuclear weapons. "The atomic bomb will be accepted far more readily if at the same time atomic energy is being used for constructive ends," a Defense Department consultant said at the time. And America became the prime promoter of both the vision and the industry throughout the world.

Decades later I was impressed by the passionate advocacy of this form of power by a few otherwise antinuclear physicists who had helped make the bombs and then devoted themselves to warning of their dangers. In that advocacy there was the hope that this dreadful weapons technology they themselves had created could somehow be turned around to serve humanity. And I suspect that people in general, without having had anything to do with creating the nuclear world-destroyer, have similar feelings. While today no one can be naïvely enthusiastic about "atoms for peace," that image of redeeming the technology still haunts us. All the more so because "humane" nuclearism is now offered as salvation from climate threat. We do have an ironic cultural comeback, however. A rock band adopted the name, yes, "Atoms for Peace" and its first album, in 2013, was called *Amok*.

A problem for climate researchers that makes their mental struggle different from that of their nuclear counterparts is that images representing global warming catastrophe can never match those of nuclear threat. True, photographs of the melting of gigantic arctic icecaps are striking and troubling—especially when associated with what has been called the "Big Melt" or "Galloping Melt" that has occurred between the years 2010 and 2015. But even these cannot compare with photographs of the destruction in Hiroshima and Nagasaki or even of nuclear testing. Nor have such climate images contributed to films with the power of *Dr. Strangelove* or *On the Beach*. We have reason to believe that visual expressions of global warming will become much more vivid and threatening as climate change progresses, but we are always likely to have more difficulty finding climate images that can approach the searing images of nuclear war—or fully evoke the civilization-destroying potential of global warming.

No wonder climate researchers and activists have frequently been mired in pessimism and even despair. They express profound doubts about what they have accomplished or can achieve in the face of inexorable climate effects. The Paris climate agreements of December 2015 have made an important dent in this pessimism, and the ramifications of Paris may go still further in helping those researchers and activists to feel less alone in their struggles. But they must now face a new burst of influence by climate rejecters spurred by the election of Donald Trump. I will discuss both the continuing hope of the Paris accord and these persistent struggles over awareness and action. But first we need to look further at the dangerous ramifications of what I call *malignant normality*, both nuclear and climate.

5

Malignant Normality

All societies construct, at various levels of consciousness, their particular versions of what is considered a normal mentality. That mentality includes ways of behaving, feeling, and thinking in significant situations, and also contains ideologies and worldviews that tend to glorify or at least look favorably on one's nation's history. The requirements of normality can be much affected by the political and military currents of a particular era.

I experienced a jolting reminder of how arbitrary societies can be in their designation of normality in an encounter I had during my study of Chinese thought reform in Hong Kong in the mid-fifties. I met with a French physician who, like many foreigners in China, had been accused of espionage and sent to a Chinese prison. There he had undergone three and a half years of a physically abusive version of "reform," during which he made a false confession of having been a spy. When I interviewed him just days after his release from the Chinese prison, he seemed anxious and very confused. He had an immediate question for me, which he asked under some pressure: "Are you standing on the People's [Chinese Communists'] side or the Imperialists' [Westerners'] side? Because from the imperialistic side we are not criminals and reeducation is a kind of compulsion [but] from the People's side it [reeducation] is to die and be born again."

He was asking about which of the two versions of normality I held to: did I believe that he had been exposed to cruel physical and psychological pressures to alter his beliefs and his identity? Or that he had been guided by beneficent reformers to a valuable personal change by means of a psychological experience of death and rebirth? But he was also, most basically, asking which of the two versions he himself now believed in.

As a psychiatrist, I have had to be concerned with the judgment of normality that is applied to distinguish those who can be diagnosed as having a psychiatric condition or disease from those who cannot. But like many psychiatrists, I have been skeptical about clear-cut contrasts between the normal and abnormal, and aware that those considered normal may be no less disturbed than those given a psychiatric diagnosis. And we have witnessed in our time considerable political manipulation of criteria for normality and pathology: for instance, the practice in the Soviet Union of diagnosing heretics with "creeping schizophrenia" and their ideas of political dissent as "delusional."

Most of the psychological interviews I have done over decades have been not with people seeking psychiatric help because of some disturbance, but rather with those I have sought out because I wanted to learn about experiences they had under certain extreme conditions. Exposed as they were to the pressures of Chinese thought reform, the grotesque effects of the atomic bomb in Hiroshima, the atrocity-producing situation in Vietnam, and the unmitigated evil of Auschwitz, my inclination was to view their *situation*, not they themselves, as abnormal. I tried to understand those extreme conditions psychologically and historically, and found that there was little to be gained by applying psychiatric categories to people caught up in them.

More than that, I came to recognize that the greatest threats to society were not posed by psychotic or severely depressed people but by those considered normal. It is the latter who participate in destructive collective projects, such as promoting and fighting wars or building, and projecting the use of, nuclear arsenals.

Nuclear and climate threats have both undergone malignant forms of normalization that suppress and distort

our perceptions of their danger. But there are contrasts between nuclear and climate normality, and those differences
tell us a great deal.

Nuclear Normality and the Logic of Madness

Hiroshima sensitized me to the absurdities of nuclear normality. After six months of living in that city and immers-
ing myself in the grotesque effects of the atomic bomb, the
embrace by anyone of the much more powerful hydrogen
bomb seemed bizarre to me in the extreme. I was most focused on my own society but strongly aware of parallel arrangements being made in the then Soviet Union. Nuclear
policies considered normal seemed to me quite insane—a
form of ethical insanity rather than a specific mental disease in the clinical sense. But that ethical insanity could
extend into collective thought and behavior. So I spoke of
the logic of nuclear war scenarios (we drop our bomb on
Moscow to stop your invasion of Europe, you drop yours
on New York, then we both stop) as "the logic of madness."

I became aware that, from the time of our testing and
use of atomic bombs, our society has attempted to render them part of the everyday landscape, viewing them as
available instruments for defending our territory and our
values, for maintaining something we call "national security." The language we have used in our arrangements for

them—"nuclear exchanges" (like gift giving), "nuclear air raid drills" (with boy scout–like participation), and digging ourselves out if we had "enough shovels" (everybody pitching in)—has served to render them manageable, domesticate them, make them ordinary and quotidian. Even in their staggering destructive power they were to become part of our normality. Which meant that, as Americans, we were being socialized to an environment dominated by them. That kind of normality can run deep in the collective national psyche.

Positive Nuclear Scenarios

One can view this normalization as occurring in waves, the first of which could be called that of *positive nuclear scenarios*. It began with President Harry Truman's official announcement of the weapon's use in Hiroshima in 1945: "The force from which the sun draws its powers has been loosed against those who brought war to the Far East." Then on a naval vessel, in a more informal statement to assembled officers and sailors, Truman declared: "This is the greatest thing in history."

That image of the weapon's world-historical "greatness" was significantly influenced by William Laurence, the journalist employed by military command to write most of the reports about the bomb. Laurence was nothing short of rhapsodic as he took on the task of becoming the bomb's spokesman. He viewed the Trinity test the month before Hiroshima as "the birth of a new world," and spoke of the bomb generally as "a gigantic Statue of Liberty, its arm raised to the sky, symbolizing the birth of a new freedom for man." Use of the weapon was meant to not only kill and destroy but also to astonish and awe; the users, however, turned out to be as astonished and awed as the victims. Laurence's descriptions suggested a mystical entity that we needed to keep among us permanently for its transcendent benefits.

The writers of positive nuclear scenarios, such as Herman Kahn and Edward Teller, expressed no such awe or mystery. They suggested that nuclear weapons, like any other, could be used to win wars, even if they resulted in casualties in the millions or hundreds of millions. Kahn suggested a dialogue in which the American president says to his advisors, "How can I go to war—almost all American cities will be destroyed?" Their answer: "That's not entirely fatal, we've built some spares," an assumption that is grandiose and borders on the delusional. Here is the extreme reach of nuclear normality.

I had no direct contact with Kahn, but there was at least one exchange during which he commented on my work. After a lecture in which he described fighting, winning, and recovering from a nuclear war, someone in the audience asked whether my account of the extent of death and suffering in Hiroshima (in my book *Death in Life: Survivors of Hiroshima*) did not cause him to reexamine some of his assumptions. His answer, as told to me by that questioner, was that my study was "soft" and appealed mainly to people's "emotions." By "soft" he meant insufficiently statistical. He was invoking a form of scientism—based on the claim to scientific purity—found all too frequently in what we call the social sciences. He was disdainfully excluding from the category of science psychological approaches that are qualitative rather than quantitative, work that includes attitudes and feelings brought out by the interview method.

A leading military figure, Rear Admiral William S. Parsons, was an equally staunch advocate of normalizing the weapon. Parsons had been a prominent member of the Manhattan Project, was the weaponeer of the *Enola Gay* responsible for the assembling of the bomb on the flight to Hiroshima, and was deputy commander of the 1948 atomic tests at the Pacific Island of Eniwetok in the Marshall Islands. Parsons was concerned that opposition to the weapon undermined American morale and made us "vulnerable to a war of nerves." His articles appeared in popular magazines including those aimed at the young, such as *Boys' Life*. He denounced the physicists' warnings about the weapon as "scientific propaganda," contributing to "a state which could well border on hysteria" and to an overall "atomic neurosis."

Edward Teller was a brilliant scientist but, as I discussed in chapter 3, one whose nuclearism could drown out his intellect when he promoted scenarios of victory in "limited" nuclear wars. He also became a self-proclaimed psychologist who denounced the "monstrous anxiety" and "psychological barrier" that interfered with "the necessary preparation for limited nuclear warfare," and urged that we revert to "rational behavior" as opposed to "anxiety . . . feelings of guilt . . . [and] fears of improbable and fantastic casualties." That is: Don't be anxious or feel guilty, don't listen to those stories of extraordinary casualties, just be steady and normal as you prepare for nuclear war.

Psychiatrists Weigh In

In fact there were real psychiatrists saying essentially the same thing. Along with other physicians and social scientists, these psychiatrists were part of a 1956 Federal Civil Defense Administration panel that prepared a report titled "The Human Effects of Nuclear Weapons Development." The panel was tasked with finding ways to cope with the "threat of annihilation" among Americans, which could undermine their willingness "to support national policies which might involve the risk of nuclear warfare." The report acknowledged that massive nuclear attack would result in "an initial shock reaction" along with fears of radioactivity, problems of bereavement, and various forms of "maladaptive" behavior. But the group insisted that it was "possible to prepare effective psychological defenses for nuclear attack" and that Americans could be enabled to accept "basic national security considerations" and reject "wild exaggerations and misinterpretations."

To achieve these goals the panel recommended an extensive therapeutic-style grassroots discussion program to be conducted in an atmosphere of "calm deliberation" without too much stress upon the kind of "awareness of annihilation" that leads to "attitudes and behavior . . . attuned to the avoidance of nuclear war, no matter what the cost." The project was to be "a monumental effort in the field of public enlightenment, formal and informal, using mass and individual media" and making use of "all educational leaders and publicists."

Chaired by a psychiatrist, this panel extended its version of nuclear normality into a call for national rejuvenation. It called upon the American "pioneer background and inheritance" that had enabled us to "count hardships a challenge" in proposing an effort at "patriotic renewal and spiritual advance." The report carried normality even beyond rejuvenation in the direction of redemption via nuclear war: "The extremity of human disaster might be the opportunity for resolute survivors" who could "draw inspiration from our forefathers and . . . make all American generations one and . . . raise hope for new dynamics of the human race." Here "humane" nuclearism, guided by psychiatrists and other luminaries, takes the form of redemptive fantasy reminiscent of the earlier Bohr conversations but exceeding them in the claim of human transformation.

President Eisenhower, who had spoken publicly about avoiding "hysteria" concerning nuclear weapons, liked the report, but its call for a national therapeutic project of patriotic awakening was more or less ignored. Nuclear normality could go only so far. The normalizing project was enhanced by the vision of "Atoms for Peace," which trumpeted the benefits of nuclear energy while deflecting some of the fear aroused by the weaponry. The atmosphere of the time was conveyed by the sequence of names considered for this key Eisenhower message—from "Operation Candor" to "Operation Wheaties" to the name that stuck, "Atoms for Peace."

Varieties of Civil Defense

What we call civil defense has been a lynchpin for this and subsequent waves of nuclear normality, but never without considerable skepticism. The shelters were mostly underground bunkers made of cement, stone, or steel, meant to protect people from nuclear fallout, and were referred to as either "nuclear air raid shelters" or "fallout shelters." They were first built extensively in Switzerland, but took hold in the United States during the Cold War. They included many kinds of structures—mainly for individual families, but also for various groups or neighborhoods, or subway stations with shelter characteristics for the general public, etc. A public opinion poll in the late 1950s showed that 40 percent of Americans were seriously considering building a shelter. Indeed the whole national project was related to the idea that either side *could* and probably *would* make use of nuclear weapons. President Jack Kennedy strongly recommended "a fallout shelter for everybody as rapidly as possible." And Wall Street investors estimated that the bomb shelter business could gross up to $20 billion in the coming years, as one recent article recounted, "if there would be coming years."

Many who built shelters were afraid that, come the time to use them, they would be invaded by neighbors who had been imprudent in their failure to provide them. And there was frequently put forward an ultimate ethical question: "Should a neighbor seek to enter your shelter and take up its valuable oxygen, is one entitled to shoot him?" That such a question could be seriously raised is an indication of how bizarre nuclear normality could become. Yet many undoubtedly perceived a considerable element of absurdity about that ethical "dilemma." The absurdity was furthered by an uncertainty about whether, when the attack came, a shelter would really protect one and one's family.

One recent retrospective look at shelters concluded: "And as the immediate terror of nuclear holocaust began to fade, Americans began to accept that fallout shelters probably did little to protect them from nuclear disaster. The backyard bomb shelters became wine cellars, fruit cellars, or just quietly filled up with water."

Involvement in civil defense was to begin early in life with the infamous "duck and cover" drills that took place during the 1950s and 1960s. In 1952, for instance, one survey found that 95 percent of elementary schools in cities of fifty thousand or more participated. Children were to duck under their desks, or cover their heads with paper, in order to protect themselves from the bomb's effects.

These drills had very mixed results, as the writer and psychologist Michael Carey discovered in a study he did in the mid-1970s of forty young men and women who, like Carey himself, had participated in them. Carey was working with me at the time and I had the opportunity to join him in conducting a few of the interviews. We were both impressed with the intensity of the psychological impact of the drills. The message children received was to suppress their fear and take on "faith [in the] ability to survive, no matter what the danger." The second part of the message was that the nuclear problem was under control, that their teachers and parents and the culture in general did not want them to be troubled by it. In fact, six-year-old kids were too perceptive to believe comfortably in either aspect of the intended message. But they were greatly troubled by the drills, and their anxiety could surface in various ways, including frightening nuclear nightmares, particularly likely to occur when there was public discussion of the possibility of the bomb's use.

I believe that Carey's study reveals much about the overall dynamic of nuclear normality: its paradoxical tendency to enhance nuclear numbing while instilling lasting anxieties that could be readily activated. Unable to believe in the message of safety and control, children were left with a sense of confusion and absurdity concerning the weapons, the drills, and the authorities who were telling them that everything was all right. Upon becoming adults, they underwent intense versions of the double life experienced by Americans in general: they knew that everything—themselves, their families and friends and teachers, all that they had ever touched or known or loved—could be extinguished in a moment. Yet they and everyone else went about business as usual and suppressed bomb-related fears in a pattern of psychic numbing that could spread to other areas.

This first wave of nuclear normality was bound up with what I call *psychism*, the claim that mental determination could overcome alternate technological destruction. The mind could be taught how to overcome fear of nuclear annihilation, how to survive a nuclear attack, and how to build new cities to replace those destroyed. The psychism was in the service of accepting the new weapons as a form of social normality, and stigmatizing opposition to them as an expression of troublesome collective neurosis.

"Owls" and Nuclear Ethics

The second wave of nuclear normality had, as its theme, *Living with Nuclear Weapons*, the title of what could be called the flagship volume of the Harvard Nuclear Study Group in the early 1980s. The message was that the weapons were here to stay, we should do our best to "reduce . . . the likelihood of [nuclear] war even though we cannot remove the possibility altogether." But "living with nuclear weapons is our only hope." Another book by the group, *Hawks, Doves, and Owls*, provided ostensible aviary wisdom: "hawks" wanted too many weapons and were too quick to use them; "doves" wanted too few weapons and absolute nonuse; and "owls"—the authors themselves— advocated caution and control in relation to the weapons but put a heavy focus on a "credible nuclear deterrent," meaning a definite willingness to use them under certain circumstances. We must, the owls told us, live with some "risk of nuclear war." Owls were wary of "overselling arms control" because that "runs the risk of lulling the public and elected officials into complacency, so that they are unwilling to invest adequately in defense," and committing oneself to drastic nuclear disarmament was "a form of atomic escapism." (Nobody seems to have told the Harvard Nuclear Study Group that owls are described by aviary authorities as "nocturnal birds of prey.")

The second wave sought to bring about a deeper integration of the weapons into American political, military, and broad social consciousness. Hence, Joseph Nye, the Harvard group's leading theorist, in a separate book called *Nuclear Ethics*, published in 1986, acknowledged that "deterrence depends upon some prospect of use." Nye added, in monumental understatement: "And use involves some risk that just war limits will not be observed," meaning that millions of noncombatants would be killed. Nye claims to ground his nuclear ethics in philosophical logic and warns against "outrage" as an "emotivist approach" that "generally prevents reasoning." A foreword to *Living with Nuclear Weapons* by Derek Bok, then president of Harvard, provided an authoritative imprimatur to the book. In it, he tells us that the book's purpose was to "inform the people" and provide a reasoned middle ground between "denying that nuclear weapons exist" and "finding refuge in simplistic, unexamined solutions."

Unrecognized here was an increasingly accepted alternative principle that true nuclear ethics would prohibit use of these genocidal devices under any conditions, and instead required a commitment to nuclear abolition, even if pursued incrementally.

This second wave of nuclear normality asks Americans to accept a more advanced form of nuclearism: a more elaborate rationale for sustaining an embrace of the weapons to deter enemies, to keep the peace, and to be used if necessary.

No wonder that the writers of *Living with Nuclear Weapons* had derisive things to say about Stanley Kubrick's classical antinuclear film *Dr. Strangelove or: How I Learned to Stop Worrying and Love the Bomb*, and Tom Lehrer's satirical antinuclear songs. Both of these expressed precisely the outrage—the "emotivist approach"—that the Harvard study group warned against but which, I insist, is completely appropriate for discussing the world-threatening danger of nuclear weapons. In *Dr. Strangelove*, an insane general launches a nuclear first strike that no one can stop and a crippled scientific advisor miraculously rises from his wheelchair declaring, "Mein Fuhrer! I can walk." The film ends with an apocalyptic montage of nuclear detonations featuring the actor Slim Pickens (a former rodeo rider) straddling a nuclear bomb and letting out a wild Texas yodel to a background recording of "We'll Meet Again (Don't know where/Don't know when)."

The *Living with Nuclear Weapons* advocates have had similar disdain for Lehrer's bitingly amusing lyrics: "So long Mom, I'm off to drop the bomb. So don't wait up for me," and "We will all go together when we go."

For such second-wave normalizers, Kubrick's film was "pure fantasy" and Lehrer's lyrics no more than the release of tension. But Kubrick and Lehrer were exposing the grotesque absurdity at the heart of nuclear normality and in the claim that "our only hope lies in living with nuclear weapons."

The Grandiose Rescue Technology

The third wave of nuclear normality, interwoven with the second, is that of the Strategic Defense Initiative (SDI), or "Star Wars." SDI depends upon a specific technology, that of antiballistic missiles, including ground-based nuclear x-ray lasers, to intercept nuclear missiles—or, as then-president Ronald Reagan put it in a 1983 speech, "to give us the means of rendering . . . nuclear weapons impotent and obsolete."

SDI can be considered a grandiose rescue technology. Such a technology is often called upon to replace human responsibility, as are grandiose schemes of geoengineering in connection with climate threat. SDI was to become part of a vast overall technological system that would include both the weapons themselves and the means of ostensibly destroying them. But SDI could never offer more than a partial defense, meaning that some nuclear weapons would inevitably reach their targets. Moreover, the enterprise was widely seen as encouraging a first-strike mentality in the possessors of SDI, who might consider themselves able to blunt a nuclear counterattack, or in an adversary to an SDI possessor, who might be tempted to embark on a "preventive" nuclear strike.

An unproven and unprovable technology, SDI was from the beginning immersed in fantasies related to earlier projections of Edward Teller that influenced Reagan and possibly to a fictional secret weapon featured in a 1940 film, *Murder in the Air*, in which Reagan appeared. So enamored was Reagan of SDI that when he and Gorbachev came close to an agreement at Reykjavik, Iceland, in 1986, to eliminate all nuclear weapons, a sticking point for the

American president was his insistence that there be no restrictions on further SDI research, a provision unacceptable to Gorbachev.

In one sense SDI could be seen as an extension of the illusions surrounding civil defense: from the insistence by T.K. Jones, a defense department official, in 1981, that in the event of nuclear attack one had to "dig a hole" and cover it over with dirt and "if there are enough shovels to go around, everyone is going to make it"; and the assumption that schoolchildren could survive such an attack by ducking under their desks or holding a piece of paper over their heads. As one skeptical observer put it, with Star Wars "the sky itself is to be converted into one vast schoolroom desk, under which we will collectively huddle while teacher hurls erasers at the marauding invaders." Nuclear normality was to be extended by including an interactive system of lethal nuclear weaponry and a visionary, fantasy-laden defense against it.

Normality and Strangeness

Nuclear normality never went uncontested. From the time of the physicists' antinuclear movement, which began even before the bomb was dropped, there have been individuals and groups seeking to expose and remedy our nuclear situation, culminating in what can be called the *nuclear swerve* of the early 1980s, which I will describe in the next chapter.

Even as nuclear normality held sway in the United States and elsewhere in the world, there were nagging voices questioning its claim and pointing to its madness. Consider, for instance, the protagonist in the 1955 film, *Record of a Living Being* (shown in the United States as *I Live in Fear*) by the great Japanese filmmaker Akira Kurosawa. The central character is a prosperous elderly factory owner who, because of Hiroshima, Bikini Atoll, and subsequent bomb tests, becomes obsessed with the danger of nuclear weapons, viewing Japan as especially vulnerable to fallout because of its special location in a kind of celestial valley. He becomes determined to move with his family to Brazil, which he believes to be safer. But his plan is contested in court by a legion of family members (including present and previous mistresses). They question his sanity, which he does indeed begin to lose, burning down his own factory because family members had focused on it as a reason he could not leave. He is eventually confined to a mental hospital and, in the last scene, he is attended to by a psychiatrist, whose words directly expose the fallacy of nuclear normality:

Whenever I see this patient, I feel very melancholy. I know that the mentally ill have a sad existence. But when I see this particular patient, I myself—though I am supposedly normal—feel quite uncertain about things, because I feel that maybe we who are able to be normal are really the strange ones.

The film did not receive a great deal of attention and was considered less successful artistically than much of Kurosawa's work. But Kurosawa's psychiatrist could be seen as a soft-spoken equivalent of Kubrick and Tom Lehrer in suggesting that "maybe we who are able to be normal are really the strange ones."

Yet nuclear normality prevailed throughout the Cold War, and dangerous elements of it persist to this day.

Climate Normality

Climate normality is equally malignant. My calling it the ultimate absurdity is a way of saying that this particular version of normality, if uninterrupted, could do us in as a species. But climate normality operates differently from its nuclear counterpart. It requires no external objects—no genocidal devices introduced by human beings for the specific purpose of killing millions of members of their own species. The problem of nuclear normality could not come into existence until the bombs were scientifically envisioned, technically constructed at Los Alamos and other centers, tested at Alamogordo, exploded in Hiroshima and Nagasaki, and then tested extensively in the Marshall Islands, other sites in the Pacific, in Australia, in Algeria, and in the Soviet Union, China, and the United States.

Climate normality, in contrast, seems to be built into our world. We experience the climate as part of "eternal nature." Actually, climate change has been no less human caused than nuclear threat; brought about by two hundred fifty years of industrialization, with dangerous spikes in recent decades. But the process has been largely silent. Only recently have we engaged in systematic study of fluctuations in planetary temperatures and in that way made ourselves aware of the menace of global warming.

Mostly a faith in climate normality has been a given for humans, for how we eat, how we heat and cool our homes, how we produce, build, and travel—for how we exist in our habitat. It has done much to define who we are and how we navigate our individual and collective lives. We are born into climate normality and have great difficulty breaking out of it.

Conspiratorial theories about climate change—Senator James Inhofe's characterization of it as "the greatest hoax" and Donald Trump's rendering it a "Chinese hoax"—are the delusional edge of a wide spectrum of disbelief and resistance. Figures like Inhofe and Trump and the Koch brothers move from that delusional edge to a political position of belligerent defense of climate normality. But they do so (as I will discuss in subsequent chapters) in the context of the climate swerve, of a vast increase in awareness of what confronts us.

Climate normality is constantly challenged by the crescendo of disasters and abnormal temperatures throughout the world, and by the consensus of the scientific community that climate change contributes to the severity of these events. Nor can we feel safe in our distance from such afflicted places as the Marshall Islands and Bangladesh; we experience the vulnerability of our own coastlines and of the planetary atmosphere we share. We do not succumb easily to the troubling paradox that the same fossil fuels that have so benefited our civilizing processes now threaten our future, yet we have no choice but to succumb to it.

The larger point is that, with climate threat, our psychological relationship to time is rapidly changing. One way of looking at this altered relationship to time is in terms of the diminishing contrast between what the ancient Greeks called *kairos* and *chronos*. *Kairos* suggests a crucial moment, an overwhelming event, in which one's actions will determine the future; *chronos* describes a more incremental sequence, that of chronological time. We have generally associated climate with *chronos*, but with increasingly frequent and greater catastrophes it veers toward *kairos*. And when some scientists insist we are approaching a point of no return, climate danger would seem to occupy the same *kairotic* realm as nuclear threat. What we can say is that climate normality covers over a threat that hovers between *chronos* and *kairos*, confusing our relationship to time and our ways of imagining the future. Holding on to climate normality becomes ever more difficult.

In one sense nuclear and climate threats are reversing themselves. Nuclear danger from the beginning evoked dreadful images of destruction, but although the danger is as great as ever it has increasingly distanced itself from our awareness. Climate danger, originally a technical matter of computer projection, has become more immediate and threatening. As the gap between mind and threatened habitat increases in the nuclear case, so it decreases in the climate situation. But these forces remain active and changeable, and as I write this, new expressions of aggressive nuclearism by Donald Trump and Vladimir Putin have rekindled nuclear anxiety that is not contained by nuclear normality.

Of course there are prominent political leaders in the United States, mostly Republicans, who still resist the actuality of climate change. Typically they move from denial and conspiracy (climate change as a "hoax") to reciting an assigned mantra, "I am not a scientist" (implying that there is scientific uncertainty about climate change)—a sequence from falsification to hypocrisy. Now that mantra is heard less frequently, probably because of the immediate evidence of global warming along with the insistent assertion of its scientific consensus. Everyone in some portion of the mind knows of the existence of climate change, but for a considerable number of people, recognizing and acting on its truths would violate their own and their sponsors' individual and group identity and ideology, especially concerning action on the part of government, regulatory arrangements, and international cooperation. Hence, what occurs now is best called *climate change rejection.*

With climate change rejection there is an insistence on holding to climate normality, and an alliance between that normality and political power. Edward Said once referred to "the normalized quiet of unseen power." Both nuclear and climate normality have served large corporate interests, so that in psychological terms our sense of what exists and is "normal" can become inseparable from our adaptation to power structures that so influence our lives.

Religious views can enter into these normalizing power structures, often with the claim that only God could alter the climate, so that whether or not it is changing has nothing to do with human action. While this view is put forth most insistently by evangelicals and biblical literalists, it can be held more widely by people who in varied ways submit themselves to larger forces, visible or invisible, and divest themselves of a sense of responsibility.

Although the climate swerve has shaken up psychological attitudes, climate normality has hardly disappeared and continues to encourage feelings of helplessness before omnipresent forces of nature. Climate normality remains unique in its total envelopment of human life and in its all-consuming threat to the human habitat. This can result in feelings of malaise and guilt over our inability to head off forces of catastrophe. Normality is perpetuated by psychic numbing that is in turn encouraged by the continuous cacophony of political falsification and rejection. But the numbing is becoming less effective in fending off the evidence, and climate normality is being increasingly questioned. Much of the questioning comes from people with special knowledge, people we call professionals.

6

Witnessing Professionals

Over the course of my work I have been interested in the behavior of professionals in the extreme situations I studied. Professionals tend to be socialized to the mores of their societies and to claims of normality. But as we have seen, there are notable exceptions who question that normality and bear witness to its dangers.

Whether doctors, lawyers, clerics, scientists, teachers, computer program designers, writers, psychologists, artists, or morticians, professionals bring a body of knowledge and skills to their work and their standing in society. We can hardly be surprised that professionals have special importance in relation to nuclear and climate threats. But before I explore that importance, I want to say something of what I learned about professionals from my work with Vietnam veterans and Nazi doctors.

In Vietnam I was struck by the malignant normality in which American doctors—I paid special attention to psychiatrists—found themselves. When an American soldier would experience anxiety or revulsion (hard to differentiate in that situation) in connection with the war, the psychiatrist or medic/assistant he was sent to would have the task of helping the soldier become strong enough to remain at duty, which often meant strong enough to participate in daily atrocities. While most psychiatrists who found themselves assigned to that role were reasonably humane people who took such matters into account and did the best they could, I found myself wondering how it was that members of my own profession could be placed in such a context.

The answers I came to had to do with the project the psychiatrists were serving, in this case the Vietnam War. They also were related to a broad sense of what a professional is and does. Looking into the history of professions in the West, I discovered that the term has religious origins and derives from a "profession" (or confession) of faith or of commitment to a religious order. But from the sixteenth century onward, with the secularization of societies, the word came to mean (according to the *Oxford English Dictionary*) "to make claim to have knowledge of an art of science" or "to declare oneself expert or proficient in a particular occupation," especially "the three learned professions of divinity, law, and medicine," along with "the military profession." Thus there was a historical shift from ultimate spiritual dedication to mastery of a specialized form of so-cially applicable knowledge and skill. While it was neces-sary to extricate the concept of the profession from simple religious affirmation, I would argue that the pendulum has swung too far in the direction of technique. What has been lost is a sense of the larger purpose of a profession and the specific ethical impact of its practitioners. The American psychiatrist in Vietnam was assigned to "help" troubled soldiers re-enter an atrocity-producing situation because he was serving military command and bringing his techni-cal skill to a profoundly immoral project.

I had in the past generally accepted psychiatry's defi-nition of itself as a healing profession with a focus on the mind. I came to recognize that a member of any profession, whatever its claim to healing, faces ethical questions about the kind of work one does, for whom, for what purpose, and about the individual and social impact of that work.

These questions were searingly raised during the Iraq War era from the 1980s through the 2000s when psychi-atrists, and psychologists even more so, lent their services to the American military in devising "enhanced interroga-tion techniques" that spilled over into systematic torture. Such procedures, some of them derived from the more bru-tal expressions of Chinese thought reform, were originally made as part of the training of soldiers to prepare them for what they might expect from an immoral enemy. Their embrace by American interrogators resulted from a decision by the highest authorities to use torture, and a willingness on the part of psychological professionals to collude in that decision and provide it with a pseudoscientific aura. In that way two notorious psychologists played a crucial role in establishing the normalization of torture.

In my study of Nazi doctors I encountered the ultimate professional surrender to malignant normality. The doctors adapted all too well to joining the Nazi project of mass murder. In Auschwitz they were actually in charge of the killing, though none of them killed anyone before or after the Auschwitz assignment. The majority of them were not extreme ideologues, but were drawn to the Nazi promise of revitalization—of making the country outwardly, and its citizens inwardly, powerful. They did not subscribe to the full Nazi biomedical vision of an originally strong Nordic race that had been "infected" by Jewish influence, which needed to be eliminated to "cure" that same Nordic race and render it strong again. But they could consider there to be a "Jewish problem" that required some form of "solution."

Many of the German doctors were appalled by what they encountered upon arriving in Auschwitz, but were then guided by old timers who drank with them and offered them a kind of perverse psychotherapy. This could include reassurance that they would get used to the place, that they would be respected if they did their job, that they were not responsible for the Jews being brought to the camp, and that they might even save a few lives during the selections process. In this way most doctors made their adaptation to killing. One could say they were *socialized to evil*—and could go about their duties, as one Jewish prisoner doctor put it, "just as someone who goes to an ordinary office to do his work." They were, that is, doing what an Auschwitz doctor was expected to do, conforming to Auschwitz normality.

It was inevitable that professionals would contribute to malignant nuclear and climate normality as well. A blatant example (mentioned in chapter 5) was the participation of psychiatrists, physicians, and social scientists in government panels seeking to help American people cope with the fear of nuclear annihilation by encouraging the desired psychological behavior for coping with nuclear attack. They were part of a larger process of pathologizing those who questioned the nuclear arms race. In the case of climate, we find less clear-cut support from professionals for malignant forms of normality. Given the strongly established scientific basis for global warming, the professional who pathologizes those who question climate normality is likely to be corrupt, a hired gun, a blind contrarian, or some combination thereof. Yet such professionals, reputable scientists among them as we have seen, have indeed emerged.

The Witnessing Professional

But professionals don't have to serve the cause of malignant normality. They are also capable of using their knowledge and technical skills to expose and bear witness to such normality, to become *witnessing professionals*. That was what I was trying to do during the Vietnam War, even though I didn't have the words at the time, by helping to organize what we called rap groups consisting of antiwar veterans and antiwar psychological professionals. The groups were initiated by the veterans themselves who wanted both psychological help from us and an outlet for making known the painful realities of the Vietnam War. That is, we were trying to help Vietnam veterans to become strong enough to carry out their antiwar function, and we ourselves, in an egalitarian setting (we did not call them therapy groups), were helped by the veterans to take in truths about the war and to find an outlet for our own strong antiwar feelings. The rap groups were limited to perhaps a few hundred people, but they did provide a model for a later outreach program of the veterans administration that came to include thousands of veterans in a setting that encouraged them to explore what happened to them in the war.

Similarly, in the doctors' antinuclear movement we could make use of our professional knowledge and authority in talking about what nuclear weapons do to cities, hospitals, and human beings. It was a form of medical, in my case medical-psychological, witness.

With climate problems, professionals of various kinds—whether engineers, physicians, or architects—are called upon to help people adapt to the extreme weather that threatens their way of life. That is crucial work of continually increasing importance. But in doing so, professionals can go beyond the problems of any particular group and become witnesses to what global warming is doing to our species.

Consider the work of Kari Marie Norgaard, an American sociologist of Norwegian extraction, who spent a year in 2000 to 2001 in a small Norwegian town to observe the attitude of its people toward the very visible effects of climate change. Although temperatures were warming and snowfall was dramatically diminishing in ways that affected everyone's life, these otherwise politically active people, who were in no way ignorant of climate change, resisted thinking and talking about it or discussing it publicly. Norgaard described their considerable use of psychic numbing and experience of a double reality of the kind I had identified in reaction to nuclear weapons. People adjusted to climate change by modifying their habits, manufacturing snow when it was insufficient for their own skiing needs and those of their tourist industry. Norgaard titled her book *Living in Denial: Climate Change, Emotions, and*

Everyday Life and has made use of her research experience in her strong public advocacy of combating global warming.

Her approach paralleled my own, going as far back as my Hiroshima study in 1962. I understood that my task in Hiroshima was to record people's experience, make known what an atomic bomb could do to a city, and, through a rigorous psychological lens, tell something of the story of what had happened to Hiroshima and its inhabitants. Over time I came to understand myself to be a witnessing professional, and the work to be an expression of *advocacy research*. This is consistent with a declaration on the part of a climate change subgroup of Psychologists for Social Responsibility: "Our problem is dedicated to understanding the psychological implications of climate change and susceptibility across the globe as we strive to influence the human behavior that has led us to the brink of severe environmental degradation."

Certainly climate scientists have been witnessing professionals, performing advocacy research since they began to sound the alarm about the dangerous effects of fossil fuels on the atmosphere and the oceans. Increasingly, they have been able to combine computer projections with direct observations, such as the melting of glaciers and the depletion of coral reefs, rendering their witness more immediate and more difficult to reject. Witnessing scientific professionals have been the intellectual vanguard of the climate swerve.

7

Climate Swerve 1:
From Experience to Ethics

What I am calling the climate swerve is a matter of individual and collective awareness. It is a state of mind and not in itself a form of action, but it can lead to action. Swerves are not orderly. This one seems particularly haphazard and, in almost all of its details, unpredictable. Yet we sense that we are in the midst of something formidable and ultimately hopeful. One thinks of Bob Dylan's insistence that "Something's happening here but you don't know what it is. Do you, Mister Jones?" With climate swerve, we all become Mister Jones in our uncertainties. But there are some observations that we can begin to make.

The term "swerve" comes to us from Lucretius, the Roman poet and philosopher who lived during the first century BCE. Lucretius was referring to a movement in the small particles he believed constituted our universe, a movement that was an unexpected deviation from the ordinary.

The contemporary humanist Stephen Greenblatt argues that the recovery in the fifteenth century of Lucretius's manuscript (after it had languished in obscure libraries for more than a millennium) contributed to a different kind of swerve: the extraordinary shift in human consciousness associated with the Renaissance and the creation of the modern world. Greenblatt himself seized upon the word and used it as the title for his compelling book, which he subtitled, "How the World Became Modern." But he was far from the only writer interested in the Lucretian swerve (or its Latin equivalent, "*clinamen*"): Jonathan Swift referred to it in 1704 in *A Tale of a Tub*; James Joyce alludes to it in the beginning words of *Finnegans Wake*; it appears in the work of Jacques Lacan and Simone de Beauvoir. And Harold Bloom wrote of the "swerve" of poets from their predecessors in *The Anxiety of Influence*. Not surprisingly these writers have different views of the swerve, but they share a sense of it as a significant, if not always logical or clear, shift in the way people experience their world.

Making Use of Death Anxiety

Greenblatt points out that Lucretius's poem is "a profound therapeutic meditation on the fear of death." Lucretius insisted that "death is nothing to us," and only by accepting its inevitability and its end to all feeling can we achieve vitality and enjoy the pleasures of life. The climate swerve is involved with death in a number of ways. It is partly a response to the fear of death, individual and collective, associated with advanced global warming. Until recently this fear has been suppressed and denied. The same has been true of the death anxiety associated with nuclear weapons, but in that case the descriptions and images of megadeaths could be more prominent and lasting. I believe that the death anxiety of climate change has moved more toward the surface as the swerve has taken hold.

Beyond climate and nuclear threats, death anxiety in general tends to be greatest when there are profound psychohistorical dislocations—breakdowns in the social arrangements that ordinarily anchor human lives. I have in mind the contentious and ambivalent relationships in our society with institutions having to do with family, religion, sexuality, birth and death, and above all with political authority and governance. Such dislocations have characterized our time, and have been greatly intensified by both nuclear and climate threats. A vicious circle of increased death anxiety, suppression and psychic numbing, and reinforcement of climate normality can result.

Yet something more constructive is also occurring. By confronting dire catastrophe and taking in the resulting death anxiety, even the possible death of our species, we make the swerve possible. And the swerve itself has an integrative effect that can, in turn, utilize the increasingly conscious death anxiety. That death anxiety, no longer avoided, becomes a stimulus for a continuous dynamic of awareness and potential action. In that way the swerve creates a state of mind appropriate to the threat. And death anxiety becomes an animating force that both enhances, and is kept in check by, the swerve.

The Evidence

The evidence of the climate swerve has three elements: experience, economics, and ethics. People throughout the world are increasingly *experiencing* the effects of climate change and the ever more active reporting of those effects in the media. None can fully avoid the drumbeat of disaster. Nor can they ignore the scientific consensus about global warming and the increasing scientific evidence that it plays an important part in the severity of hurricanes and tornadoes, extreme droughts and wildfires, heat waves, periodic moments of extreme cold, rising sea level, and extreme flooding. All this is part of the growing recognition that climate danger encompasses our planet and that we are all vulnerable.

There has been a quantum leap in media coverage, which I confirmed by scrutinizing the articles about global warming in the *New York Times* during the years 2013, 2014, and

2015. There is not only greater coverage of climate disasters in general, as compared to past years, but a much greater tendency to raise questions about the role of global warming in causing or intensifying them.

Public opinion polls and attitude studies show that increasing numbers of people are becoming convinced that the earth has been dangerously warming and that human influence has been a crucial factor. Even the disingenuous assertion "I'm not a scientist" is becoming less useful as a political talking point, though it is sometimes replaced by "There has always been global warming," which is a way of expressing doubt about human causation. Or a still more sophisticated evasiveness, as expressed by Scott Pruitt, who President Trump appointed to head the Environmental Protection Agency: "Science tells us the climate is changing and human activity in some manner impacts that change. The ability to measure and pursue the degree and the extent of that impact and what to do about it are subject to continuing debate and dialogue."

Now human influence is admitted but there is still an insistence upon ultimate uncertainty and doubt. Denial and falsification have by no means disappeared (Pruitt has discounted the danger of carbon dioxide emissions), and political opposition has in some ways intensified. But the typical oppositional stance has become a combination of partial, mostly unspoken, recognition of climate change as an entity, and rejection of climate change as a significant threat requiring social and political action.

Such rejection is necessary to sustain one's identity and worldview, including antagonism to government regulations and, in many cases, to government and governance. Yet climate rejecters are increasingly on the defensive and subject to the disapproval of the general public. To be sure, their focus on lost jobs and Third World development can be politically compelling. And they can rightly point to technical problems in a massive conversion to renewable energy sources. But in doing so they cannot help revealing the extraordinary technical and economic revolution that is taking place in the production and utilization of those renewable sources and their potential human benefits.

This contradictory position of rejecters can lead to extreme actions, as in the behavior toward scientists by the Trump transition team in late 2016. The questionnaires they sent to the Department of Energy asking for the names of employees (scientists and staffers) and information about their attendance at climate change meetings suggested a McCarthyite witch hunt and potential purge. Climate deniers and rejecters have long expressed this antiscience stance, which is both anti-Enlightenment and antimodern. It is also a manifestation of the kind of totalism I could observe in Chinese communist thought reform: subsuming the truth to a closed, falsified narrative, while seeking to control the media and other sources of information. Significantly, the questionnaires were withdrawn in response to expressions of outrage that could be understood as opposition to totalistic witch hunts and, at least indirectly, as affirmation of the climate swerve.

Risky Business and the Rockefeller Turn

One powerful economic indication of the swerve was expressed in the emergence in 2014 of a group called Risky Business, which published a national report titled "The Economic Risks of Climate Change in the United States." Leaders of the group include two former secretaries of the treasury (Henry Paulson and Robert Rubin), a former secretary of state (George Shultz), and two prominent American billionaires (Michael Bloomberg and Tom Steyer). What is notable is that these financial tycoons focus specifically on the economic costs of climate change. As made clear in their mission statement: "The signature effects of human-induced climate change—rising seas, increased damage from storm surge, more frequent bouts of extreme heat—all have specific measurable impacts on our nation's current assets and ongoing economic activity." Their message is that our present use of fossil fuels will soon bring down the American economy. They are moneymen who are responding to the ever more pressing climate truths.

But when Risky Business seeks to "frame climate change as an economic issue," as the *Wall Street Journal* declared, it is confronting all of American business with experiential truth and also taking the issue into an ethical realm.

Still more striking was the announcement in 2014 by members of the Rockefeller family that one of their foundations, the Rockefeller Brothers Fund, would divest its holdings from fossil fuel companies. These Rockefeller family members seized upon the irony of their action, since the family fortune is derived mostly from oil, and the major fossil fuel corporations in our society (Exxon, Amoco, and Chevron) are spin-offs from the original Rockefeller Standard Oil conglomerate. The divestment statement stresses the fund's longstanding commitment to "mitigate the effects of climate change," but the fund's president made clear that the action was also based on a bottom-line calculation very much in the family tradition: "John D. Rockefeller, the founder of Standard Oil, moved America out of whale oil and into petroleum. We are quite convinced that if he were alive today, as an astute businessman looking out to the future, he would be moving out of fossil fuels and investing in clean, renewable energy."

Two years later came further, more decisive action by members of the Rockefeller family, this time as leaders in exposing and condemning the fraudulent behavior of

ExxonMobil, the most gargantuan corporate descendant of their oil baron Rockefeller forebear. Now the Rockefeller Family Fund joined the Rockefeller Brothers Fund in divesting from fossil fuels and both groups gave financial support to an investigative journalism project (carried out by Columbia University graduate students and by the website InsideClimateNews.org) on the overall behavior of Exxon over the past forty years.

This led ExxonMobil to accuse the Rockefeller groups of "funding a conspiracy" against them. David Kaiser, the president of the Rockefeller Family Fund and a fourth-generation Rockefeller himself, responded with a hard-hitting two-part essay, written with Lee Wasserman, exposing ExxonMobil's duplicity and probable fraudulence in relation to climate change. The oil conglomerate was a corporate leader in state-of-the-art climate research in the late 1970s and early 1980s, but suddenly reversed itself in 1988 and began to support front groups for climate denial and the scientists working with them (the scientists could be hired guns or ideologues of fierce nationalism—some were veterans of Cold War nuclearism—and intense opposition to government regulations). The essay referred to ExxonMobil's "tobacco strategy," as revealed by Naomi Oreskes and Eric Conway in *Merchants of Doubt*, the systematic effort of the tobacco industry to counter scientific findings that their product was a cause of cancer and other fatal illnesses. "Doubt is our product," as one spokesman put it.

Kaiser and Wasserman described how Exxon sponsored some of the same corrupt scientists who had done the dirty work for the tobacco industry to claim "scientific uncertainty" where there was none. The essay pointed out that the CEO of Exxon began to make false statements—claiming in 1997 that "the earth is cooler today than it was twenty years ago." The corporation also began to use its considerable political influence to press the George W. Bush administration to rid itself of scientists who insisted on climate truths and replace them with climate falsifiers. When a number of attorneys general from major states—New York, Massachusetts, and California—later initiated investigations of ExxonMobil for defrauding investors and the public, the Rockefeller Family Fund informed them of what they had learned about the corporation's behavior. The essay minced no words: "For over a quarter-century the company [ExxonMobil] tried to deceive policymakers and the public about the realities of climate change, protecting its profits at the cost of immense damage to life on this planet."

In this way Rockefeller family members have emerged as leading voices in scrupulously documented confrontation with the largest direct corporate descendants of

John D. Rockefeller's original Standard Oil conglomerate. The essay and the committed behavior of these Rockefeller family members and foundations suggest a powerful turning point in the climate swerve and its spreading awareness. In an interview Wasserman also spoke of what he called "the obvious historical irony of the fact that we are Rockefellers doing this."

Congressional climate rejecters have mounted a bullying antiscience response, defending the ExxonMobil accusation of "conspiracy" and issuing subpoenas to the Rockefeller groups and other climate activists. The Trump transition team also became indirectly involved via the Breitbart News Network (formerly headed by Steve Bannon, Trump's leading advisor), the alt-right website featuring racist and conspiratorial material, which called for an investigation of the environmental groups for conspiracy. The battle between truth-telling Rockefellers and climate rejecters, including Republican congressmen, the Exxon-Mobil Corporation, and right-wing websites, will undoubtedly continue indefinitely, but it will take place in a world increasingly conscious of climate danger.

Another significant economic contribution to the swerve was the formation in 2016 of a coalition of powerful industrial leaders and consultants who formed the Task Force on Climate-related Financial Disclosures. Its purpose is to make available reliable disclosures about climate issues to "lenders, insurers, and investors"—that is, to tell it like it is about the dangers of global warming for specific economic enterprises. Task Force members come from the world's twenty richest nations, and the group has held plenary meetings in London, Singapore, Tokyo, Washington, D.C., and Paris, where it has raised such questions as how corporations would fare under the two degrees Celsius warming limit laid out in the Paris climate accord, including effects of extreme weather, rising seas, and market and currency swings. The members of the Task Force do not see themselves as altruists so much as pragmatically self-interested businessmen. They have responded to pressures from corporate shareholders for accurate information about dangers to their investments. As the chair of the larger group, Michael Bloomberg, put it: "Climate change is not only an environmental problem, but a business one as well." The more general point is that the largest business enterprises—the giant corporations—are ultimately no less vulnerable to the climate threat than the rest of us.

Stranding the Assets

Economics and ethics also converge in the revelatory term "stranded assets." The term was suggested by a group called Carbon Tracker Initiative, composed of financial and energy authorities dedicated to analyzing carbon investment risk. Fossil fuel assets are underground sources that can potentially be extracted and utilized. They become "stranded" when significant impediments to that extraction arise. The term "stranded assets," originally used in connection with stock exchange maneuvers, came to suggest a pivotal climate truth: extracting a significant amount of those underground fossil fuel assets would have catastrophic effects on the human future. To avoid that outcome, Carbon Tracker has estimated that 70 to 80 percent of those assets must not be used, in other words, they must remain "stranded."

Psychologically speaking, to be stranded is to be separated from the safety of human contact, from fruition or realization. One speaks of being stranded in a completely inaccessible place—at sea, or in the desert. There is the example from baseball of stranded runners, those left on base and unable to come "home." In the case of fossil fuels, the assets that *need* to be stranded—kept from fruition or realization, kept separate from us as human beings—have long been a source of all-purpose energy and wealth but have become a species-threatening poison.

Stranded assets is about money, and the value of the fossil fuel assets that need to be left in the ground to avoid planetary catastrophe is estimated to be in the area of

$20 trillion. That means that fossil fuel corporations and their national partners need to surrender that staggering amount of untapped wealth in order for the planet to avoid reaching the two degree Celsius threshold of catastrophe.

Historians and economists tell us that the last time in American history that such extraordinarily valuable stranded assets existed was in 1865 and the "assets" took the form of human beings. Slaves then constituted 50 percent of the Southern economy, 16 percent of the overall American economy, and by today's standards added up to a monetary value of about $10 trillion. That was the value of the assets abolitionists were demanding that slave owners forfeit, the amount that would be stranded when America freed itself of its most evil institution. Following the cotton boom, slavery was never more valuable and never more threatened—precisely the situation in which fossil fuel corporations find themselves today. It took the bloodiest war in American history, and something on the order of seven hundred thousand deaths (the historian Drew Gilpin Faust called her unsparing narrative *This Republic of Suffering: Death and the American Civil War* for good reason), for the radical demand of abolitionists to be realized. The lesson: people don't give up financially valuable assets easily.

Our question now is, can we keep stranded that which would otherwise do us in? I find myself comparing the situation to an old Jack Benny joke, in which an armed robber offers the comedian a choice, "Your money or your life!" There is a brief silence and Benny then says, "I'm thinking it over." Like Jack Benny, America and the world are "thinking it over."

Some of these reflections are far from life-enhancing. There are arguments made, arguments claiming to be ethical, for converting all fossil fuel assets, leaving absolutely none of them "stranded." That claim, made by a few fossil fuel CEOs and their defenders, goes something like this: By leaving no asset behind we carry through our fiduciary responsibility to stockholders—and there could actually be significant financial return in the short term, even if oil stocks, in the long run, can be seen as problematic. Also, by going ahead with the excavation of these stranded assets, corporations and their officers can claim the virtue of attending to the energy hungers of developing countries and thereby helping millions of people to emerge from poverty. And there is the further claim of sustaining jobs and buttressing overall economies that benefit from coal (despite its economic demise), oil, and gas extraction. In the past that perspective could have been viewed as practical and even ethical.

Stranded Ethics?

These arguments for species-threatening poison take us into a realm we may call *stranded ethics*. With the established truth of global warming there is need to leave such arguments in the ground, keep them stranded. In April 2014 the CEO of ExxonMobil was guilty of stranded ethics when he reiterated the longstanding corporate principle of pressing ahead with maximum extraction of fossil fuels from all available sources. At the same time he insisted that the corporation "vigorously" consider the risk of climate change in its planning while "meeting the economic needs of people around the world in a safe and environmentally responsible manner." Interestingly, he too played with the word "stranded," insisting that by calling forth all of its assets, ExxonMobil is "preventing [worldwide] consumers—especially those in the least developed and most vulnerable economies—from themselves becoming stranded in the global pursuit of higher living standards and greater economic opportunity." There is no mention of the necessary world shift to renewable energy sources.

As we have seen in the case of ExxonMobil, corporate leaders were faced with scientific findings that questioned the overall morality of their companies' activities, which were revealed to contribute to a staggering threat to human beings in general. Such corporations either had to cease these activities or else claim human value for them, such as that of contributing to a better life for millions of people in developing countries, in order to blunt the scientific truths. Here we may speak of an ultimate expression of stranded ethics—a corporation's assertive justification of activities that threaten the human future.

Haunting the entire ethical discussion—what we may call ExxonMobil's "Jack Benny delusion"—is that previously quoted sentence from Kaiser and Wasserman's essay on behalf of the Rockefeller Family Fund: "For over a quarter-century the company [ExxonMobil] tried to deceive policy makers and the public about the realities of climate change, protecting its profits at the cost of immense damage to life on this planet." Yet this convergence of experience, economics, and ethics could bring hope for a new outcome. It is expensive to extract fossil fuels from the ground, even using advanced methods for extraction known as fracking. In the face of the climate swerve with its dissemination of awareness, and of the pledges of governments to restrict carbon emissions, those corporations may view the considerable cost of the extraction of fossil fuels from the earth as a bad investment.

In fact, it is we human beings in general who would be stranded by unmitigated climate change—stranded as a species from our own habitat. In that sense, the passions of the swerve are directed at staying connected to our source, and at taking actions that allow us to be no longer stranded from it.

There was a parallel demand expressed frequently in the physicians' antinuclear movement in the 1980s. It took the form of a toast (often at night after a bit of social imbibing) made by either a Soviet or an American doctor: "I drink to your health, and to that of your people, and also your leaders. Because if you survive, we survive. If you die, we die." Underneath its gallows humor, the toast was a demand that nuclear protagonists emerge from their own stranded ethics of potentially world-ending conflict and recognize the truth of our shared fate. The sense of the toast is readily translatable into climate terms, where the claim of beneficent use of fossil fuels is replaced by a climate swerve also built around the recognition of shared planetary fate. That recognition is expressed in a question posed by the writer and speaker Duane Elgin: "When will humanity express its moral outrage that it is wrong to devastate an entire planet for countless generations to come, just to satisfy the consumer desires of a fraction of humanity for a single lifetime?" Elgin's question is now frequently asked and the "moral outrage" he seeks is very much the stuff of the climate swerve.

Florida and the Unsayable

While the swerve in no way dominates political life in America, politicians are nonetheless affected by it. A sequence of events in Florida illustrates both resistance to the swerve and the increasingly shaky ground of that resistance. In 2011, Rick Scott, a longstanding denier of climate change, was elected governor of Florida, a state extraordinarily vulnerable to the effects of global warming. His method for maintaining this contradiction was that of embracing the Republican mantra, "I am not a scientist." While that mantra has helped avoid speaking a truth one is partly aware of, its ethics is on the order of a person answering "I am not a criminologist" when asked whether he or she is opposed to murder.

But various state agencies inevitably became involved with Florida's all too overt problems of eroding coastlines and coral reefs and flooding of coastal areas. At the same time, as a lawyer involved with these agencies put it, state employees were given "a warning to beware of the words global warming, climate change, and sea-level rise." Later "sea-level rise" was reinstated as usable "because for some projects it had to be taken into consideration," as one former official explained. And an angry University of Miami professor declared that "it will be hard to plan for climate change . . . if officials can't talk about climate change."

Scott would later deny any such prohibition of these words, but when pressed by journalists to say whether or not he and his department believed in global warming, he avoided an answer and instead declared, "I'm into solutions, and that's what we're going to do." He pointed to the state's "significant investments in beach renourishment [and] flood mitigation." Those efforts, according to knowledgeable observers, were highly inadequate. Perhaps they had to be because, as that University of Miami professor said, it was hard to plan for climate change if officials couldn't talk about climate change. Putting things another way, appropriately strong efforts at climate mitigation require acceptance and articulation of climate truth. Our human minds must be given full rein if we are to preserve our habitat, or at least prevent further catastrophic damage to it.

That Florida scenario suggests the unstable mental equilibrium around climate rejection, which can exist between holding on to one's political identity—as antigovernment, antiregulation, anti–public works, business and private sector oriented, and sensitive to large funders who deny or reject climate change—and the ever more insistent truths about global warming. Florida is an especially intense example of that conflict in awareness because so many of its residents live on coasts in which rising seas are so common as to cause what has been called "sunny-day flooding."

South Florida in particular has been considered the ground zero for sea-level rise.

Sufficiently alarmed by Hurricane Matthew in 2016, Governor Scott urged people to seek safety away from the coast, declaring, "This storm will kill you!" But he never uttered words like "climate change" or "global warming" or even "sea-level rise." Significant in the evolving awareness of climate threat in the state was the exposure of Governor Scott's contradictions by such reputable groups as the Florida Center for Investigative Reporting, the *Miami Herald*, the Union of Concerned Scientists, and various scientists working in Florida. Terms such as "bordering on criminal negligence," "criminal," and "Orwellian nightmare" were used.

We have come to use the term "Orwellian" to indicate language that Orwell sometimes referred to as "doublespeak" or "doublethink"—language that obscures or reverses the truth. Scott's behavior suggests something on the order of "nonspeak" or "nonthink" in the sense of sustaining a taboo against words such as "global warming," words that convey climate truths. The concept of taboo has always conveyed vehement prohibition of a statement or action that violates the deepest, most sacred elements of a culture, or in this case a subculture that is clinging to its antigovernment, antiregulatory principles.

Ten Florida scientists wrote to Donald Trump in December 2016, just before he took office, asking for a meeting with him to explain the perils of rising sea levels, pointing out that "much of your Mar-a-Lago Club [his opulent vacation center in Palm Beach] could be under water in coming years because of climate change." And fellow governor Jerry Brown of California expressed similar clarity in anticipation of a visit by Scott to his state, declaring: "If you're truly serious about Florida's economic wellbeing, it's time to stop the silly political stunts and start doing something about climate change—two words you won't even let state officials say. The threat is real and so too will be devastating impacts."

Brown was giving expression to regional climate policies undertaken by states and municipalities. Brown's state of California aims to reduce its carbon emissions by 40 percent below 1990 levels by 2030, with an emphasis on Green Enterprise Zones, extensive planting of trees, and embracing of renewable energy sources. Many urban leaders in the United States have accepted a German concept of a "regenerative city," one that remains aware of its natural ecosystem and regenerates the resources it utilizes, and there is worldwide discussion of a "new urban agenda" that focuses on cities' minimizing harm to the environment and combatting global warming. A writer on the subject, Jeff Biggers, takes us beyond cities when he speaks of "wind turbines that rise out of the cornfields" and clean energy across the heartland.

Scott and other "I-am-not-a-scientist" political leaders can vary in their actual convictions, but we may suspect that all of them, to some degree, both believe and do not believe in climate change: they have some degree of recognition of its actuality while rejecting the belief itself because it would require too much of a change in their political identity, worldview, and affiliations. The question here is how long this unstable truth-rejecting equilibrium can hold, how long it can be sustained in the face of evidence at one's immediate coast, or front door, evidence of the actuality that Scott and others like him reject. The beginning answer is already apparent in polls showing an increase in the number of Republicans who believe in climate change and in the formation of Republican groups whose purpose is to increase acceptance of climate change. This evolving awareness can be suppressed by hardline falsifiers and rejecters, but only partially and temporarily.

The unstable mental equilibrium concerning climate change denial extends to the highest levels, including Donald Trump and his appointees and followers. Even as he rejects climate change as a problem that is human caused, Trump has applied for permission for a sea wall to protect his elite golf course in Scotland from flooding associated with climate change. And he is surely aware of the warnings by Florida scientists of the danger posed by rising sea levels to his beloved Mar-a-Lago resort.

Environmentalism and "Essential Facts"

Climate deniers and rejecters can hide behind a longstanding concern about "nature," a movement of environmentalism that long preceded the recognition of global warming. Though the term "environmentalism" was coined in 1922, its roots are in nineteenth-century struggles against the harmful environmental effects of the industrial revolution. William Wordsworth's embrace of the British Lake District as a "sort of national property, in which every man has a right and interest who has an eye to perceive and a heart to enjoy" suggests the connection with literary Romanticism and "a return to nature." Environmentalism has been embraced by people who wish to protect nature, for example beautiful coastlines such as those in Florida. This has enabled some to claim a love of nature while ignoring or rejecting the idea of climate change. But it is also true that the environmentalist movement has been a passionate source of climate activism. Indeed, one aspect of the climate swerve is widespread love of the natural world and realization that we are destroying it. Thus, when Governor Scott was given, by his own administration, an award for his "environmentalism," that award became a source of ridicule and was viewed as part of the Orwellian climate situation in the state.

In these matters Henry David Thoreau was a remarkable voice of truth. Thoreau used his deep connection with the natural environment as a passionate critique of much of modern life. He has become something of a patron saint for climate activists, and justly so. His quest in the Walden woods was for what he called "the essential facts of life," by which he meant the most basic requirements (or "necessaries") of human existence. By exposing excesses in consumption and exploitation, lives of too much "getting and spending," Thoreau was setting the ground for what we now call sustainability. He also speaks to climate activism in his declaration: "Rather than love, than money, than fame, give me truth." His truth required constant vigilance against deceivers and against self-deception. He acted on these convictions by combining his environmentalism with disobedience to unjust laws. He insisted that telling the truth about how things are would be the end of slavery in America. Mohandas Gandhi was surely expressing his debt to Thoreau, which he fully acknowledged, when he spoke of "experiments with truth." And Václav Havel's commitment to "living in truth" called forth a similar spirit. This principle of truth telling, together with life-enhancing activism, is at the heart of the climate swerve.

The Pollution Connection

Further evidence of the connection between general environmental concerns and the climate swerve is awareness of the pollution of the environment, perceived as poisoning, as causing what I called in my Hiroshima work a fear of invisible contamination. I have encountered a similar fear in people exposed to toxic substances leaked into the environment. In the 1980s, the sociologist Kai Erikson and I served as consultants in class-action legal cases, interviewing plaintiffs. We studied people who had been exposed to large gasoline spills in two Colorado cities, one in Northglenn involving the Chevron Corporation, a successor company to Standard Oil, and the other the Royal Petroleum Corporation, a smaller company with connections in the western part of the country. People spoke of their fear of a poison that they could not see or otherwise detect, of being vulnerable to "serious diseases down the road," especially some form of cancer, emphasizing that "you don't know, and the unknown is what bothers you."

In Hiroshima, as we have seen, the fear of invisible contamination was based on the reality of delayed radiation effects, which could bring about fatal forms of leukemia and other cancers. Of course the situation in Hiroshima was infinitely more extreme than that faced by people in Colorado exposed to gasoline fumes. But those fumes have their own dangers, acute and delayed, including carbon monoxide poisoning and harm to the lungs. Whether the feared invisible contamination is related to radiation or gasoline, there is a sense that the poisons let loose in the external environment had entered your own body, poisons that could strike you down, perhaps kill you, at any time, without warning.

There is evidence that pollution was present in earliest recorded history, or even prehistory. Recent studies suggest the existence of extensive soot on the ceilings of ancient caves, and of pollutants in preserved Paleolithic teeth that are four hundred thousand years old. More elaborate evidence dates urban pollution back over centuries, so that by the second half of the nineteenth century there were pollution problems in many cities in Europe and the United States, which in turn led to movements to bring about cleaner air. Such response to pollution is by no means the same as later recognition of climate change. But longstanding problems with pollution have contributed greatly to our sensitivity to our more recent confrontation with global warming.

Pollution and Nuclear Fallout—Rachel Carson

The American biologist and writer Rachel Carson, working in the mid-twentieth century, became a key historical figure through her simultaneous focus on the pesticide DDT and the fallout from nuclear testing as dangerous forms of pollution. Her classic volume, *Silent Spring*, published in 1962, reactivated the environmentalist movement and took it in the direction of combatting hidden pollutants. Significantly, DDT had been considered a great protector of life in its widespread use as an insecticide effective in the control of malaria and typhus. Carson wrote lyrically about the herbicide's role in the death of animals and birds, to the point of endangering America's symbolically treasured bald eagle.

Carson's ability to articulate close connections between pesticides and radiological fallout was striking. Introducing her subject in the first pages of *Silent Spring*, she spoke of the pesticide chemicals as "sinister and little-recognized partners of radiation in changing the very nature of the world—the very nature of its life." Her outrage connected with, indeed was inseparable from, passionate concerns at the time about the dangers of nuclear testing and particularly of Strontium-90.

Strontium-90, released through nuclear explosions into the air, comes to earth in rain or drifts down as fallout, lodges in soil, enters into the grass or corn or wheat grown there, and in time takes up its abode in the bones of a human being, there to remain until his death. Similarly, chemicals sprayed on croplands or forests or gardens lie long in soil, entering into living organisms, passing from one to another in a chain of poisoning and death.

Carson, as her biographer emphasizes, reflected intensifying Cold War concerns about nuclear fallout, including warnings by scientists and physicians that it could result in leukemia, bone cancer, and genetic alterations. Indeed, throughout her work, radiation and chemical defoliants were identified as the "two products of wartime science . . . forever linked in discovery, destruction, and debate."

Moreover, as leading environmental historian Ralph H. Lutts noted, "These [nuclear] pollutants, particularly fallout, played a special role in preparing the public for Rachel Carson's message." Or as another historian put it: "Anybody who has been alarmed by atmospheric pollution from nuclear tests could see that [Rachel Carson] was talking about other dimensions of the same problem." Lutts wrote a detailed article entitled "Chemical Fallout: Rachel Carson's *Silent Spring*, Radioactive Fallout and the Environmental Movement." Overall, he observed, "She and her book were products and representatives of their time, as well as shapers of it."

Rather than using the word "insecticides," Carson preferred the term "biocides," which anticipated the later use of the term "ecocide." And when she referred to the new historical moment in which "One species—man [has] acquired significant power to alter the nature of his world," she anticipated the concept of the Anthropocene epoch to describe our present era, dominated as it is by baleful human influence. Carson was moderate in her opinions and favored the continued use of limited amounts of DDT for malaria prevention. She was nonetheless attacked by chemical corporations who called her "a fanatic defender of the cult of the balance of nature."

But some of the most intemperate and inaccurate attacks on her have been posthumous, made by contemporary groups who falsify and deny climate change. Those who engaged in this retrospective attack, as Oreskes and Conway tell us, "didn't just deny the facts of science. They denied the facts of history."

Carson was a natural target because her work encompassed nuclear and climate dangers. More than anyone else, she threatened the identity of both nuclearists and polluters, and she expressed her truths in words that people could understand (*Silent Spring* has sold more than 2 million copies). The climate swerve has a major resource in Rachel Carson's work.

Health and Illness

Pollution affects health and arouses fear of illness. Studies have shown that people are more drawn to climate change issues when these are presented as threatening their own health and that of their children. In April 2015, President Barack Obama held a roundtable at Howard University, joined by the U.S. surgeon general and the head of the Environmental Protection Agency, and chaired by the prominent CNN doctor, Sanjay Gupta. Vivek Murthy, the surgeon general, emphasized the connection between climate change and asthma and referred to a study by the American Thoracic Society that implicated climate change as a cause of extensive lung disease. Obama mentioned development of new apps by Google and Microsoft that would enable people to monitor the immediate air quality in their communities. And the director of the World Health Organization referred to the staggering statistic of 7 million people killed in 2012 by exposure to indoor and outdoor pollutants.

The effort here was to connect climate with pollution and with immediate concerns about illness and health in one's own neighborhood and one's own family. As *Time* magazine suggested in 2013, "Medical professionals may be the best messengers for global warming." One might well substitute "witnesses" for "messengers." But the larger point is that imagery of environmental poisons that threaten people's health increasingly feeds and prods the climate swerve.

It turns out that Tom Lehrer had something to say

on these matters. He provided what could be called pollution-and-health lyrics, sung in calypso style:

If you visit American city
You will find it very pretty
Just two things of which you must beware
Don't drink the water and don't breathe the air

And, similarly:

See the halibuts and the sturgeons
Being wiped out by detergeons....
Just go out for a breath of air
And you will be ready for Medicare

We would do well to send Tom Lehrer on a world goodwill and deep-breathing tour, with mandatory stops in Beijing and Los Angeles.

Swerve and Ecocide

As noted earlier, the term "ecocide" has been lodged in my mind since I heard Galston introduce it in 1970 at the Congressional Conference on War and the American Conscience. This was followed by Richard Falk's efforts, over the years, to bring international legal authority to combatting ecocide. I also associate the term with my own study of Hiroshima and its radioactive poisoning as the "ultimate pollution."

Ecocide—its danger and our struggle against it—is at the heart of the climate swerve. There is the disturbing sense, on the part of increasing numbers of people, that we humans are committing ecocide through our poisonous carbon and methane emissions into the atmosphere and the oceans. In the act of destroying the habitat of human beings and other animal species, we are threatening the habitat of forests and plants and botanical life in general.

In recent years the British lawyer Polly Higgins has mounted an energetic campaign combatting ecocide by developing "laws and governance to prevent the destruction of our planet." For the same purpose the University of London Human Rights Consortium has developed an "ecocide project." Related to all this is an extraordinary book by the science writer Elizabeth Kolbert. In *The Sixth Extinction*, Kolbert records the five great extinctions thought by scientists to have occurred on our planet in the past and asks whether we are now undergoing a sixth, distinct in being human caused. Such immersion, both thoughtful and dire, into a narrative of species extinction is a necessary prod to the climate swerve.

Our awareness of the danger of species extinction is subject to ebbs and flows. But it has reached a point of recognizing the planetary reverberations of local climate actions, a globalization of consequences. When Republican senators and congressmen in one way or another block fossil fuel restrictions, the results affect not just Americans but Indonesians and Nigerians. Similarly, when the Chinese permit Beijing to be overcome with pollution, much of it related to fossil fuels, they are affecting residents of, say, Marseilles and New York City. And if the British vote to leave the European Union means that they engage less cooperatively with EU members on climate issues, the manifestations are likely to be felt everywhere. Similarly, when Los Angeles takes steps to reduce global warming, benefits, even limited ones, are experienced in Kyoto. That is, climate change and the swerve in awareness are part of a global dynamic of quick and powerful mutual influence.

8

Climate Swerve 2:
Awareness and Adaptation

Freud was skeptical about our capacity for individual or collective autonomy. He understood us to be creatures of instincts, or drives, which tended to overwhelm most claims of conscious agency. Yet toward the end of his life he declared: "The voice of the intellect is a soft one, but it does not rest until it has gained a hearing." And he viewed this insistence as "one of the few points on which one may be optimistic about the future of mankind."

Formed Awareness

Psychologists since then have given more attention to parts of the psyche in more regular touch with the outer world—to the ego, the self, or the sense of identity. But there has been little focus on what I would call the general human capacity for awareness. Awareness includes two interwoven levels of experience. One has to do with being watchful or vigilant (related to words like "wary," "beware," and "guard"). The other has to do with being actively cognizant and conscious (related to words like "expectation," "seeing," and "awe"). Overall, awareness suggests intense interaction with the environment on the part of the mind informed by potential threat.

When exploring responses to nuclear threat, I came to make a distinction between *fragmentary awareness* and *formed awareness*, a distinction which holds true for climate threat as well. Fragmentary awareness consists of images of climate change that are recurrent but fleeting. These may include pictures in our heads of a typhoon in the Philippines, Hurricane Sandy closer to home, and of such future events as islands sinking into the sea and violent conflict of various kinds involving climate refugees and inhabitants of countries where they seek refuge. The images come and go, and may be combined or endlessly recast, but they lack clarity and coherence and can readily give way to more immediate claims on one's attention and to the numbing of climate normality.

Formed awareness, in contrast, brings such images into a more coherent narrative with elements of causation and of consequences. Media projections of extreme weather, of droughts and flooded coastlines, come together in ways that call into question previous faith in climate normality. The self, individually and collectively, can no longer suppress or reject the actuality of climate images and the considerable unease and fear that those images evoke. The unease and fear are appropriate and can be useful, all the more so as the self is liberated from convoluted efforts at falsification and denial. Formed awareness does not guarantee climate wisdom, but is necessary to it. Formed awareness, when widespread, becomes part of a social dynamic built upon climate truths, and a basis for constructive action.

Stranded Adaptation

Formed awareness is inseparable from the crucial issue of adaptation. Climate normality has provided a feeling of adaptation with arrangements that, until recently, seemed to work well enough. We humans excavated fossil fuels to warm us in the winter and cool us in summer, and to drive the technology and industry that have improved our lives and enabled us to take control of the planet. But our evolving knowledge of global warming made it clear that the seeming normality of our fossil fuels–driven world had become dangerously maladaptive. Such maladaptation is experienced socially and ecologically, and can be understood in relation to the overall evolutionary process. New kinds of adaptation are required for sustaining the life of the human species. That is what the swerve is all about. And the Paris accords of December 2015 extended the swerve to a new level of species awareness.

Just as I characterized fossil fuel CEOs as demonstrating *stranded ethics* when they insisted that their assets be excavated and used, we may also speak of their *stranded adaptation*, which has become in practice a highly dangerous maladaptation. Nor are those attitudes confined to fossil fuel company CEOs and their political allies. All of us are so accustomed to a fossil fuel economy that no one may be entirely free of impulses toward such stranded adaptation—which is why we require the most extensive kind of collective effort to overcome it.

Part of the new adaptation required is to climate change itself. I spoke disdainfully in chapter 5 of "living with nuclear weapons," since these genocidal devices can and should be disarmed and removed from their present stockpiles. (Even though the capacity to build new nuclear weapons would still exist, eliminating the ones we have would make the world a great deal safer.)

But climate change permits no such abolition—though we should get as close to eliminating fossil fuels as possible—and in any case we are already experiencing significant effects of global warming worldwide. We have no choice but to seek to survive in an ever more damaged habitat, even as we find ways to mitigate still more dangerous developments. Nor can such adaptation be adequately made while rejecting the truth of global warming, as we saw in the case of Governor Scott of Florida.

A Single Threatened Species

In December 2015, representatives of close to two hundred nations, just about all that qualify as such, came together at the Paris Climate Conference in what could be called an expression of species unanimity. Though all but a handful made pledges of reduced emissions, the most important outcome of the meeting may well have been enhanced awareness that we are all members of a single, threatened species. In that sense the essential agreement in Paris was about a cast of mind, one of great significance in its potential for contributing to *species adaptation* via climate wisdom.

Two earlier climate statements were crucial to the Paris agreement. One of them was the Pope's encyclical of 2015, directed not only at all Catholics but at "every person living on this planet," in calling for nothing less than universal "ecological conversion." Pope Francis referred constantly to our "common home," a phrase originally used by his namesake Saint Francis of Assisi, and one that has striking parallels to the nuclear quest for "common security," the principle that an adversary's security is necessary to one's own. This of course was the principle expressed by that nuclear weapons toast by American and Soviet doctors.

Can we imagine an equivalent climate toast made by Chinese or American physicians (or physicists or poets)? "I drink to your health and to the health of your people and your leaders. If you thrive by reducing significantly your pollutions and fossil fuel emissions, we thrive as well. And if you increase these pollutions and emissions, and suffer and die, then we also suffer and die."

While we can hardly count on any such toast, its spirit was expressed in the 2014 treaty between China and the United States to mutually reduce carbon emissions. That treaty might have been even more important to the subsequent Paris accords than the Pope's encyclical. Here were the two greatest polluters, the United States in the past and present and China in the present and immediate future, coming to a common realization that each had to take action on behalf of the human habitat. Limited as the treaty was, it had profound import as a suggestion of the kind of species awareness that was later manifested in Paris. Significantly, in September 2016, the two countries jointly announced their ratification of the Paris agreement, thereby assuring that it would soon be rendered active. A month later the United States and China again took the initiative with the Kigali Amendment, involving

170 countries, to radically reduce and eventually eliminate the production and consumption of hydrofluorocarbons, a particularly carbon-dangerous substance used widely in air conditioning.

The Prior Nuclear Swerve

We can learn more of the swerve dynamic by looking back at the nuclear weapons sequence. I have in mind the anti-nuclear swerve that culminated in the march and rally of one million people in Central Park in New York City on June 12, 1982. What turned out to be the largest protest event in American history was timed to coordinate with a special United Nations Session on Disarmament. The nuclear freeze campaign, based on Randall Forsberg's call for a bilateral freeze in the reproduction and stockpiling of nuclear weapons, played a central role, as did the National Committee for a Sane Nuclear Policy (or SANE).

I shared the ideas and emotions of the nuclear swerve by participating in the international physicians' movement, both nationally (Physicians for Social Responsibility or PSR) and internationally (International Physicians for the Prevention of Nuclear War or IPPNW). As doctors we could speak with authority about what nuclear weapons do to human beings and about our inability as a profession to function in any kind of recovery from a nuclear war. In that historical moment, people were ready to listen, and the IPPNW was awarded a Nobel Peace Prize in 1985.

There were also more radical protests, such as those of the Catholic left, led by the priests Philip and Daniel Berrigan, who engaged in repeated expressions of civil disobedience, such as breaking into nuclear weapons assembly plants to make powerful symbolic gestures of hammering or pouring blood on the nosecones of the weapons. My occasional testimony (usually but not always limited by judges to sentencing hearings) provided me with a connection, however tangential, to this portion of the movement. It was part of my larger involvement with an informal group of writers and activists who advocated abolition of the weapons, a position given particularly eloquent expression by Jonathan Schell. As in any such swerve, these advocacies and emotions also took hold in various ways in large numbers of people who gave them little public expression.

For this antinuclear sentiment to take shape there had to be first a dissemination and widespread recognition of nuclear truth: the capacity of the weapons to annihilate us as a species and destroy much of our planet. Some of us in the antinuclear movement insisted that it was erroneous to talk about "weapons" or about "war" and more accurate to speak of a "nuclear end." Taking in such nuclear truth required a breakout from large-scale psychic numbing. People throughout the world protested the maneuvers and threats by the American and Soviet political leaders and came to recognize the apocalyptic dimensions of destruction that would result from the use of these nuclear devices. Much of the passion and theory of the nuclear swerve came from below, that is, from nongovernmental antinuclear movements made up of ordinary people. But international diplomats associated with the United Nations, and some national officials, were also of enormous importance. There was an ever-increasing commitment, though often unspoken, to the continuity of human life.

To be sure, that swerve hardly eliminated the danger of nuclear weapons. Indeed, their very existence in the world maintains that danger, all the more so because of proliferation (the increasing number of countries possessing them), technological miniaturization (smaller weapons deemed more usable), and the failure of nuclear powers to reduce their stockpiles significantly or to cease modernizing the nuclear devices. Yet the worldwide nuclear swerve of the early and mid-eighties could well have contributed significantly to nuclear restraint, to the nonuse of the weapons since Nagasaki on August 9, 1945.

That swerve of deepened nuclear awareness could also have been important to the emergence of a world movement to eliminate nuclear weapons, exemplified by the International Campaign to Abolish Nuclear Weapons, and by a decision of the United Nations General Assembly to convene a "United Nations Conference to negotiate a legally binding instrument to prohibit nuclear weapons, leading towards their total elimination." That was to be achieved in two meetings, the first in late March 2017 and the second in mid-June of that year. Much of the energy for this new wave of abolition has come from non–nuclear weapons possessing countries (the nuclear powers have tended to oppose it), from nongovernmental organizations, and from an articulate Hiroshima survivor named Setsuko Thurlow.

Like swerves in general, this antinuclear one has had ebbs and flows of awareness. The death anxiety associated with nuclear awareness can readily give way to numbing or even nuclearism. But an actual or potential awareness of nuclear danger remains an important factor in restraint.

Light from Dark Visions

The apocalyptic twins—climate and nuclear threats—can enhance our awareness of our status as a single threatened species. When International Physicians for the Prevention of Nuclear War was awarded a Nobel Peace Prize in 1985, we had delegates from about sixty countries, as compared to the approximately two hundred countries that joined the climate accords in Paris in 2015. An important difference between the two movements is the involvement in Paris of official delegations representing nation-states (although supported by many nongovernmental organizations), as opposed to the unofficial activities of the IPPNW. But even at Paris there was little in the way of enforceable commitment and a great deal in the way of species awareness. Although a few nations at Paris declined to make pledges of reduced emissions because they believed that the richer countries should take on larger responsibility, everybody signed on to a recognition that, with the threat to our common home, none of us is safe and everyone must act. As a step forward in a universal mental state, the meetings and accords were an expression of our great evolutionary asset, the human mind, on behalf of species survival.

Underneath the champagne sipped by the Paris signatories was a dark vision of massive death and violence. Heat waves, droughts, glacial melting, landslides, and widespread flooding were becoming both commonplace and ever more menacing. By mid-June of the following year heat records were shattered across the southwestern United States, affecting such cities as Tucson, Phoenix, Las Vegas, and San Diego. At the same time observers noted a "grim new low" for [arctic] sea ice cover with the vanishing, over thirty years, of an area twice the size of Texas. And the melting of glaciers in Alaska resulted in an enormous landslide that spread rocky debris more than six miles; scientists predicted that such slides would continue to occur as warming temperatures caused greater glacial melt. And the spring of 2016 brought unprecedented flooding and damage throughout most of Europe, including Paris, resulting in the temporary closing of perhaps the world's most famous museum, the Louvre. Then in early January 2017 the National Oceanic and Atmospheric Administration (NOAA) and a number of other climate monitoring groups confirmed the fact that 2016 had registered the highest temperature on record, and that for the third year in a row a new planetary heat record was set.

The dark vision of climate death and suffering is reminiscent of the vision of nuclear Armageddon experienced by Eugene Rabinowitch in Chicago just before the testing of the first atomic bomb. In both cases imagining massive destruction and death was a prerequisite for wisdom. The threats remain, but the wisdom—the formed awareness of what those threats constitute—propels resistance to the use and even existence of nuclear weapons, and to the catastrophic impact of our carbon and methane emissions on the human habitat.

Always and Never Too Late

These dark visions cast us in the role of potential survivors who, characteristically, seek some kind of meaning in their ordeal. One such meaning is that it is already too late for us: climate change is rampant, irreversible, and more powerful than any antidotes we may bring to it. A recent book by the noted philosopher Dale Jamieson, *Reason in a Dark Time*, has a subtitle: *Why the Struggle Against Climate Change Failed—and What It Means for Our Future*. Jamieson speaks with the authority of one who has explored climate issues for decades, and his book is knowledgeable and challenging. In it, he announces the demise of "the hope that people of the world would solve the problem of climate change through a transformation in global values." (The book was published a little over a year before Paris.) Jamieson adds: "Rather than a global deal rooted in a conception of global justice, climate policy for the foreseeable future will largely reflect the motley collection of policies and practices adopted by particular countries." Here Jamieson's melancholy perspective could be consistent with a skeptical view of what happened in Paris based on the uncertainty about legal obligations. But Jamieson surely underestimated the kind of collective survivor imagination that was also expressed in Paris.

That survivor meaning has to do with the imagined death of a species—our own—and the need to mobilize its intellect and passion on behalf of its continuous life. In that process we reassert our existence as cultural animals with awareness of those who have gone before, and those who will go on after, our own brief lifespan. The Paris accords were about a broadly human struggle for a significant shift in adaptation to our habitat, for preserving the life of humankind and other species.

We cannot say how successful we will be in living out such species-oriented adaptation. We will not be able to prevent a good deal of very harmful climate change that is either already manifest or soon to strike us. But we have a genuine capacity for preventing the most extreme forms of catastrophe. That is why, from now on, our actions are always and never too late. And since failure to pursue such adaptation will surely bring catastrophe, we have little choice but to try.

Jamieson tells us that "Climate change threatens a great deal but it does not touch what ultimately makes our lives worth living: the activities we engage in that are in accordance with our values." There may be an admirable emphasis here on living as fully and ethically as we can in our era of climate change. But the statement becomes sadly absurd in light of the cataclysmic possibilities affecting all of us that could be brought about by further increase in temperatures and in ocean acidification.

The Adaptation Paradox

At issue is the kind of adaptation we make. If we consider species-wide agreement to significantly mitigate global warming to be a lost cause, our efforts at adaptation would become exclusively focused on preparations for ever more lethal climate onslaughts. This would take us into the realm I described as stranded adaptation and could even be seen as a high-toned version of climate normality. We do better to focus on species adaptation via significant mitigation of climate change, while at the same time taking necessary measures to protect individual people and whole populations from existing and anticipated climate disaster.

In advocating species awareness, I am not abandoning psychology for biology. Rather, I am insisting that the human species has become our operative group, and that our urgent need is to bring our psychological and historical imagination to the task of preserving that all-inclusive group. In earlier work, I referred to our broader ecosystem as "the habitat of all species" and quoted Archibald MacLeish in depicting human beings as "riders on the earth together." Confronting the full danger of nuclear and climate devastation enables us to sustain, rather than destroy, our species.

Evolution, Identity, and the Self

To do that we can hardly count upon an automatic form of evolutionary adaptation. But we can find some hope in what the distinguished biologist and evolutionary theorist

E.O. Wilson calls "evolution for the good of the group." I refer to a newly revived emphasis on the part of Wilson and others away from the "selfish gene" (meaning survival of those genetically related) to the survival of larger groups. Wilson pointed out that we humans "are exquisitely well adapted to live on this particular planet," but that "we must understand ourselves as a species." One thinks also of Erik Erikson, perhaps the most distinguished psychoanalytic figure since Freud, and his emphasis on ever-expanding identity. In his focus on continuity and change, Erikson spoke of the broadening of group ties, of our capacity for a "universal" or "species-wide" identity. He referred, somewhat obscurely but evocatively, to "belief in the species." Erikson was certainly bringing his identity theory into the area of what I have been calling species awareness.

In my own earlier work on nuclear weapons I spoke of an evolving sense of self that, while holding to immediate identifications with family, group, and nation, comes to identify with the entire human species. When thinking in those terms, the *species self* is inseparable from the climate swerve. We take on a species connection, so that even in our ordinary life we come to experience our symbolic immortality by living on not only in our children and their children or in our group or nation, but by living on in humankind.

The Possibilities of Proteanism

Helping us toward the expansion of awareness and identity that characterizes the climate swerve in general is a capacity of the individual self for mutability and change. I speak of this capacity as *proteanism*, in reference to Proteus, the Greek sea god and notorious shape-shifter. Proteanism includes not only a potential for change and transformation but also for containing odd combinations—seemingly unlikely or even antithetical elements—that encourage highly particular views of family and religion to combine with a new embrace of species concerns.

Proteanism is greatly influenced by historical currents. It thrived collectively during the Renaissance and Enlightenment in the West, and the Meiji Restoration in Japan. During such intense historical transitions, widespread psychohistorical dislocation is accompanied by equally widespread protean experiment, resulting in new individual and collective convictions and configurations.

We humans have an inherent potential for proteanism that has to do with our being inveterate symbolizers in the sense described by the philosophers Ernst Cassirer and Susanne Langer. They taught us that, in our mental life, with the help of our uniquely large frontal lobes, we humans take in nothing nakedly, but must alter and re-create all that we perceive. In that way our minds are continuously active, and our only way of engaging the world is to bring past experience and present inclination to everything we perceive. Or to put it another way, perception is inseparable from re-creation and change.

To be sure, there can be individual and collective forces that react to such collective proteanism by pressing to hold it in check or even to shut it down. These can take the form of widespread psychic numbing. Or the form of dogmatic reactionary and fundamentalist behavior that closes down change.

This infinite capacity for—indeed requirement of—symbolization is basic to the human mind and its relation to our habitat. It is a key to our extraordinary adaptive skills. It is the source of our ability to imagine our way beyond immediate constraints and threats. It enables us to draw upon past experience to project a different future. Hence Cassirer could call us the "symbolic animal" whose symbolizing capacity "transforms the whole of human life [so that] man lives not merely in a broader reality—he lives, so to speak, in a new *dimension* of reality."

Symbolization can be understood as a connecting point between evolution and history. It holds out an ever-present possibility for a historical surge of proteanism and social change. The climate swerve we are undergoing is such a moment.

Other threats could result in survivor-influenced proteanism, as I observed during trips to Czechoslovakia in 1990 and 1991. I found that the Velvet Revolution in that country against its oppressive Communist government released tendencies among nonviolent dissidents that they had previously been unaware of possessing. Havel himself, an extraordinary combination of writer and political rebel, expressed a highly imaginative, yet tough-minded form of proteanism in his willingness to give his life for that previously noted principle of "living in truth." Living in truth included, for Havel and many others, a capacity to alter one's behavior and take on tasks one had never attempted or known that one could attempt.

It was Susanne Langer who told us that "the organism does what it can." But there are moments when the organism—individually and collectively—does more than it thought it could. This may be our situation now, as we struggle with our swerve toward collective survival.

Prospective Survivors

The swerve can encounter impairments in the form of Trump-like rejection of climate change. It can encounter darkness as well as light. But we have seen the powerful contributions that can be made to the swerve by the imagination of catastrophe and the sense of being prospective survivors. A survivor is one who has encountered, been exposed to, or witnessed death and has himself or herself remained alive. A prospective survivor, without that immediate exposure to death, is one who imagines such a death encounter. This includes what I call the survivor's lasting "death imprint," and "image of ultimate horror," which in its extremity comes to represent the full destructiveness and pain of the deadly event. I think again of Eugene Rabinowitch walking through the streets of Chicago and imagining the collapse of the skyscrapers around him, and then finding immediate meaning in what he had imagined by intensifying his efforts to prevent the bomb's use.

To avoid actual catastrophe requires that ever-increasing numbers of us take on that sense of being prospective survivors. As prospective survivors we can find meaning in our actions to combat climate change. We can take on a survivor mission of preserving our habitat and embracing genuine forms of adaptation for our species. In doing so we reassert our larger human connectedness, our bond with our species.

I conclude with both dread and hope. Terrible world-ending threats, nuclear and climate, awaken us to our shared destiny as members of a single, overarching group, the human species. That group is no longer a distant, theoretical entity, an abstraction, but engages our minds with palpable consequences.

We can take some pride in a human evolutionary narrative of stunning adaptive skills. But we cannot hide from a sense of ourselves as a talented species in deep trouble. Can we once more call upon—and expand—these adaptive skills? Such adaptation requires translating species awareness into behavior that is species wide and species sustaining.

There has already been more suffering from climate change than we have allowed ourselves to recognize, and that suffering will increase. But we are capable of averting a civilization-ending catastrophe, and even of achieving a new beginning for our species and many others. We are capable of applying the human mind to preserving our habitat, and our own continuity of life.

Why does all this matter to a ninety-year-old man who will not himself experience the worst climate disasters that might await our species? For me it is simply a matter of that larger human connectedness. Whatever our age, we are part of a bond much greater than ourselves, part of a flow of endless generations that include forbears as well as children and grandchildren. The bond is not only biological but is related to all we do and experience in the world. This principle of the great chain of being—and I speak as a secular person—takes on special importance as we approach the end of individual life. The human chain has never been more vulnerable. Nor have we ever been more aware of the mind's capacity for attending to our species by renewing and enhancing our habitat. Of course it is very late in the game, but at the same time far from too late.

Acknowledgments

Slim volumes, in their compression of ideas, demand clarity and priority. Charles Strozier, Jane Isay, James Carroll, and Nancy Rosenblum, after careful readings of the manuscript, addressed both of these requirements in ways that much improved this book. I benefited also from exchanges with members of the Wellfleet Psychohistory Group, including Edwin Matthews, Robert Holt, and Peter Balakian. Richard Morris of Janklow and Nesbit has been a wise agent and friend from the book's beginnings. Carolyn Mugar combined research support with warm personal encouragement. At The New Press, Carl Bromley's thoughtful enthusiasm included important suggestions both imaginative and practical, about the manuscript itself as well as its worldly connections. Jed Bickman provided useful ideas about context. And Emily Albarillo offered a steady hand in turning the manuscript into a book.

Notes

Preface

ix interviewing survivors there: Robert Jay Lifton, *Death in Life: Survivors of Hiroshima* (Chapel Hill: University of North Carolina Press [1968], 1991).

x   other studies I have done: Robert Jay Lifton, *Thought Reform and the Psychology of Totalism: A Study of "Brainwashing" in China* (Chapel Hill: University of North Carolina Press [1961], 1989); *Home from the War: Learning from Vietnam Veterans* (New York: Other Press [1973], 2005); and *The Nazi Doctors: Medical Killing and the Psychology of Genocide* (New York: Basic Books [1986], 2017).

xi Paris Climate Conference in December 2015: Coral Davenport, et al., "Inside the Paris Climate Deal," *New York Times*, December 12, 2015.

1: The Ultimate Absurdity

4   "The creature who could clothe himself": Loren Eiseley, "Man: The Lethal Factor," *American Scientist* 51 (March 1, 1953): 71–83.

4   the wiring of our brains: see for example R. Gifford, "The Dragons of Inaction: Psychological Barriers That Limit Climate Change Mitigation and Adaptation," *American Psychologist* 66, no. 4 (May–June 2011): 290–302; and G.T. Gardner and P.C. Stern, *Environmental Problems and Human Behavior* (New York: Pearson, 2002). These studies emphasize difficulties, but do not suggest impossibility.

Notes

2: Hiroshima as Pollution

10 "Hiroshima and the Ultimate Pollution": Robert Jay Lifton, notes, 1962–63, box 95, Robert Jay Lifton Papers, Manuscripts and Archives Division, New York Public Library.

12 The term "ecocide" was first used: Arthur Galston, "Technology and American Power" panel, Congressional Conference on War and National Responsibility, Washington, D.C., February 1970; and David Ziegler, *The Invention of Genocide* (Athens: University of Georgia Press, 2001), 19.

13 "provided the first modern case": Richard Falk, *A Global Approach to National Policy* (Cambridge, MA: Harvard University Press, 1975).

13 Ecocide Project: Human Rights Consortium, School of Advanced Study, University of London, 2012–14.

3: Apocalyptic Twins: Nuclear and Climate Threats

19 "an act of atonement" . . . "It was necessary": M. Susan Lindee, *Suffering Made Real: American Science and the Survivors at Hiroshima* (Chicago: University of Chicago Press, 1994).

21 "within two years Reynolds had gone": Kristin Grabarek, "On the Cutting Edge: The Peace Activism of Earle Reynolds," *Friends Journal*, April 1, 2009.

22 The founding mission: See Frank Zelko, *Make It a Green Peace! The Rise of Environmentalism* (New York: Oxford University Press, 2013).

23 a new emphasis: see Ira Helfand, M.D., "Nuclear Famine: Two Billion People as Risk?," International Physicians for the Prevention of Nuclear War, Second Edition, 2013.

24 "the people of Utrik see themselves as": Kai Erikson and Robert Jay Lifton, "Sociocultural and Psychological Impacts of the Bravo Nuclear Test on the People of Utrik" (in monograph prepared for the Nuclear Claims Tribunal, Republic of the Marshall Islands, June 28, 2002).

25 "As a result of sea level rise": Michael B. Gerrard and Gregory E. Wannier, eds., *Threatened Island Nations: Legal Implications of Rising Seas and a Changing Climate* (Cambridge: Cambridge University Press, 2015).

25 "does not meet American standards": Michael Gerrard, "A Pacific Isle, Radioactive and Forgotten," *New York Times*, December 3, 2014.

26 "For almost seventy years": Phillip Muller, "Pacific Islands Face a Deadly Threat from Climate Change," *Wall Street Journal*, May 30, 2013.

26 unprecedentedly extreme rainfall and multiple landslides . . . "climate change and urban resilience": Andrew DeWit, "Hiroshima's Disaster, Climate Crisis, and the Future of the Resilient City," *Asia-Pacific Journal* 12, no. 35 (August 29, 2014).

26 the basis of his pioneering book: Michael B. Gerrard, ed., *Threatened Island Nations: Legal Implications of Rising Seas and a Changing Climate* (Cambridge: Cambridge University Press, 2013).

26 Marshall Islands leaders mounted: International Court of Justice Press Release, 2014/18, no. 25, April 2015.

27 "new politics of nuclear disarmament": Avner Cohen and Lily Vaccaro, "The Import of the Marshall Islands Nuclear Lawsuit," *Bulletin of the Atomic Scientists*, May 6, 2014.

27 their specific effects on the environment . . . "the world itself became the laboratory": Joseph Masco, *The Theater of Operations: National Security Affect from the Cold War to the War on Terror* (Durham, NC: Duke University Press, 2014).

29 referred to as "pollution": Stanley Kramer, director, *On the Beach* (Beverly Hills, CA: United Artists, 1959).

29 radiological contamination, biological weapons, weather control: Jacob Darwin Hamblin, *Arming Mother Nature: The Birth of Catastrophic Environmentalism* (Oxford: Oxford University Press, 2013), 4.

32 "if we stop" . . . "Radiation from test fallout": see Edward Teller with Allen Brown, *The Legacy of Hiroshima* (New York: Praeger, 1975), 180–81.

33 "Society's emissions of carbon dioxide" . . . a grandiose form of geo-engineering: Edward Teller, "Sunscreen for Planet Earth," *Hoover Digest* (Hoover Institution, Stanford, CA), no. 1 (1998).

33 "somewhat larger risks" . . . " 'Pull yourself together and get to work' ": see Herman Kahn, *On Thermonuclear War* (Princeton, NJ: Princeton University Press, 1959).

34 "If present trends continue": see Gerald O. Barney, study director, "The Global 2000 Report to the President" (Washington, DC: Government Printing Office, 1980).

34 "less crowded ... less polluted": see Herman Kahn and Julian Simon, *The Resourceful Earth: A Response to Global 2000* (New York: Basil Blackwell, Inc., 1984).

35 "You see what happened to me": Richard Feynman quoted in obituary by James Gleick, *New York Times*, February 17, 1988.

38 whether God would want to impose such suffering: see Charles Strozier, *Apocalypse: On the Psychology of Fundamentalism in America* (Eugene, OR: Wipf & Stock Publishers [1994], 2002).

38 In my work in the late 1990s: see Robert Jay Lifton, *Destroying the World to Save It: Aum Shinrikō?, Apocalyptic Violence, and the New Global Terrorism* (New York: Metropolitan Books [1999], 2000).

39 "forcing the end": Gershom Scholem, *The Messianic Idea in Judaism* (New York: Schocken, 1971), 56–57.

39 The Islamic State (or ISIS) also envisions: see William McCants, *The ISIS Apocalypse: The History, Strategy, and Doomsday Vision of the Islamic State* (London: MacMillan, 2016).

40 the "smaller apocalypse": see Stan Cox and Paul Cox, *How the World Breaks: Life in Catastrophe's Path, from the Caribbean to Siberia* (New York, The New Press, 2016).

41 "It was the end of time": see Robert Jay Lifton and Eric Olsen, "The Human Meaning of Total Disaster," *Psychiatry* 39, no. 1 (1976); and Kai T. Erikson, *Everything in Its Path: Destruction of Community in the Buffalo Creek Flood* (New York: Simon & Schuster, 1978).

4: Different Mental Struggles: Nuclear and Climate Truths

48 "endless talk ... deep happiness": Freeman Dyson, *Disturbing the Universe* (New York: Harper and Row, 1979).

48 Enrico Fermi offered ... just-in-case public statements: see Richard Rhodes, *The Making of the Atomic Bomb* (New York: Simon and Schuster [1986], 2012).

49 "It will hardly be possible": see Martin J. Sherwin, *A World Destroyed: Hiroshima and Its Legacies*, (Palo Alto, CA: Stanford University Press, 2003), 118.

49 influenced by Niels Bohr: see Rhodes, *The Making of the Atomic Bomb*.

51 told of walking through the streets: see Eugene Rabinowitch, "Five Years After," in M. Grodzins and E. Rabinowitch, eds., *The Atomic Age* (New York: Basic Books, 1963).

53 traditional scientific caution: Naomi Oreskes, "Beyond the Ivory Tower: The Scientific Consensus on Climate Change," *Science* 306, no. 5702 (2004): 1,686.

54 called forth extensive data: Dr. James Hansen, director of NASA Goddard Institute, in a statement to U.S. Senate Committee on Energy and Natural Resources, June 24, 1988.

55 model created by tobacco companies: Naomi Oreskes and Erik Conway, *Merchants of Doubt: How a Handful of Scientists Obscured the Truth on Issues from Tobacco Smoke to Global Warming* (New York: Bloomsbury USA, 2010).

56 declarations that "Carbon dioxide is green!": see for example H. Leighton Steward, *Fire, Ice and Paradise* (Bloomington, IN: Author-House Self Publishing, 2009).

56 "the scientists and politicians" . . . "the non-climactic effects": Freeman Dyson, foreword to *Carbon Dioxide: The Good News* (GWPF Report 18) by Indur M. Goklany (London: The Global Warming Policy Foundation, 2015).

58 overstates nuclear safety: see for example Sharon Squassoni, "The Incredible Shrinking Nuclear Offset to Climate Change," *Bulletin of the Atomic Scientists* 73, no. 1; Joe Romm, "Why James Hansen Is Wrong About Nuclear Power," ThinkProgress.org, January 7, 2016; and Naomi Oreskes, "There Is a New Form of Climate Denialism to Look Out for—So Don't Celebrate Yet" (opinion), *The Guardian*, December 16, 2015.

59 1,656 "indirect deaths": "Fukushima Stress Deaths Top 3/11 Toll: Uncertainties amid Nuclear Crisis Acutely Felt by Elderly," *Japan Times*, February 20, 2014.

59 an estimated 100,000 to 160,000 people . . . spilled more toxic nuclear waste: see International Atomic Energy Agency, "The Fukushima Daiichi Accident: Report by the Director General" GC[59]/14 (2014); and Madhusree Mukerjee, "Five Years Later, the

Fukushima Nuclear Disaster Site Continues to Spill Waste," *Scientific American*, March 1, 2016.

59 misinformation and cover-up: Daniel Kaufmann and Veronika Penciakova, "Japan's Triple Disaster: Governance and the Earthquake, Tsunami and Nuclear Crises," Brookings Institution, March 16, 2011.

59 worst nuclear power plant disaster: see United Nations Scientific Committee (UNSCEAR), "UNSCEAR Report on the Effects of Atomic Radiation Sources and Effects of Ionizing Radiation 2008 Report to the General Assembly," 2008.

59 number of cancer deaths: see Lisbeth Gronlund, "How Many Cancers Did Chernobyl Really Cause?," *All Things Nuclear* (blog), Union of Concerned Scientists, April 17, 2011.

59 by Greenpeace at as many as 200,000: see "The Chernobyl Catastrophe: Consequences on Human Health," Greenpeace website, April 18, 2006.

59 "zone of alienation": Andrew Kramer, "Chernobyl's Silent Exclusion Zone (Except for the Logging)," *New York Times*, April 23, 2016.

59 unsafe for human life: Story Hinckley, "Chernobyl Will Be Uninhabitable for at Least 3,000 Years, Say Nuclear Experts" *Christian Science Monitor*, April 24, 2016. Article also mentions estimate of 20,000 years: "Asked when the reactor site would again become inhabitable, Ihor Gramotkin, director of the Chernobyl power plant, replies 'At least 20,000 years.'"

60 "That's the same thing as what happened over in Japan": from psychological report by Kai Erikson and Robert Jay Lifton prepared for Levin, Fishbein, Sedran & Berman, Three Mile Island litigation.

62 "the industry and the government responded swiftly": Nuclear Energy Institute, "Lessons From the 1979 Accident at Three Mile Island: Fact Sheet," October 2014.

62 fifty-six significant nuclear accidents: "A Brief History of Nuclear Accidents Worldwide," Union of Concerned Scientists website.

63 twelve significant nuclear accidents in France . . . deep economic trouble: "Coverup at French Nuclear Supplier Sparks Global Review," *Wall Street Journal*; and "France's Nuclear-Energy Champion Is in Turmoil: Electricité de France Has Had to Shut Down 18 of Its 58 Nuclear Reactors," *The Economist*, December 1, 2016.

64 "The atomic bomb will be accepted": Peter Kuznick, "Japan's Nuclear History in Perspective: Eisenhower and Atoms for War and Peace," *Bulletin of the Atomic Scientists*, April 13, 2011.

65 those of nuclear threat: see Spencer W. Weart, *The Rise of Nuclear Fear* (Cambridge: Harvard University Press, 2013).

65 "Big Melt" or "Galloping Melt": see Carolyn Gramling, "Antarctic Ice Shelf Being Eaten Away by Sea," *Science*, December 4, 2014; and "West Antarctic Melt Rate Has Tripled" (joint press release), American Geophysical Union and National Aeronautics and Space Administration (NASA), December 2, 2014.

5: Malignant Normality

67 a jolting reminder: Lifton, *Thought Reform and the Psychology of Totalism*.

71 "The force from which the sun": "Top Secret Interest to Genera Graves," Manhattan Engineering, District Papers, Modern Military Branch, National Archives; timeline at nucleartimeline.org.

71 "This is the greatest thing in history": Harry S. Truman, *Memoirs, Volume I, Year of Decisions* (Garden City: Doubleday, 1955), 421.

72 "the birth of a new world": William L. Laurence, *Men and Atoms: The Discovery, the Uses, and the Future of Atomic Energy* (New York: Simon and Schuster, 1959), 116–19.

72 "How can I go to war": Kahn, *On Thermonuclear War*, 642.

73 "vulnerable to a war of nerves": Paul Boyer, *By the Bomb's Early Light: American Thought and Culture at the Dawn of the Atomic Age* (Chapel Hill: The University of North Carolina Press, 1994), 316–17.

74 "monstrous anxiety": Teller with Brown, *The Legacy of Hiroshima*, 288–89.

74 "threat of annihilation": "The Human Effects of Nuclear Weapons Development," Report to the President and National Security Council, November 1, 1956.

76 "Operation Candor" to "Operation Wheaties" to the name that stuck, "Atoms for Peace": see also Eisenhower Institute, "Eisenhower Institute Atoms for Peace + 50," October 22, 2003, Dwight D. Eisenhower Presidential Library.

77 "a fallout shelter for everybody": President Jack Kennedy quoted by

Pat Zacharias in "When Bomb Shelters Were All the Rage," *Detroit News*, July 14, 2014.

78 one survey found: Zacharias, "When Bomb Shelters Were All the Rage."

79 "faith [in the] ability to survive": for Michael Carey study, see his "Psychological Fallout," *Bulletin of the Atomic Scientists* 38 (January 1982), 20–24.

80 we should do our best to "reduce": Albert Carnesale, et al., *Living with Nuclear Weapons* (Cambridge, MA: Harvard University Press, 1983).

80 "hawks" wanted too many weapons: see Graham T. Allison, et al., eds., *Hawks, Doves, and Owls: An Agenda for Avoiding Nuclear War* (New York: W.W. Norton & Company, 1986).

81 "deterrence depends upon some prospect of use": see Joseph Nye Jr., *Nuclear Ethics* (New York: Simon and Schuster, 1988).

83 "Mein Fuhrer! I can walk": from Stanley Kubrick, director and co-writer, *Dr. Strangelove or: How I Learned to Love the Bomb* (Culver City, CA: Columbia Pictures, 1964).

84 "to give us the means of rendering": Ronald Reagan, archives, Address to the Nation on Defense and National Security, March 23, 1983, Ronald Reagan Presidential Library.

84 a sticking point: see Nickolai Sokov, "The Reykjavik Summit: The Legacy and a Lesson for the Future," Nuclear Threat Initiative, 2007.

85 "the sky itself is to be converted": see Boyer, *By the Bomb's Early Light*, 316–17.

86 "Whenever I see this patient": Akira Kurosawa, *Record of a Living Being (I Live in Fear)* (Tokyo: Toho Studios, 1955).

6: Witnessing Professionals

94 When an American soldier would experience anxiety or revulsion: see Lifton, *Home from the War*, 336–37 and 414–23.

97 "just as someone who goes to an ordinary office": see Lifton, *The Nazi Doctors*, 193.

99 use of psychic numbing: Kari Marie Norgaard, *Living in Denial: Climate Change, Emotions, and Everyday Life* (Cambridge, MA: MIT Press, 2011).

Notes

7: Climate Swerve 1: From Experience to Ethics

101 extraordinary shift in human consciousness: Stephen Greenblatt, *The Swerve: How the World Became Modern*, (New York: W.W. Norton & Company, 2012).

101 far from the only writer: see among others Mladen Dolar, "Tyche, clinamen, den," *Continental Philosophy* 239, no. 46 (August 2, 2013), 223; Sean Braune, "From Lucretian Atomic Theory to Joycean Etymic Theory," *Journal of Modern Literature* 33, no. 4 (Summer 2010), 167–81; and Michael Dirda's review of Stephen Greenblatt's *The Swerve, Washington Post*, September 21, 2011.

101 "death is nothing to us": Greenblatt, *The Swerve*, 3.

105 Public opinion polls: Richard Wike, "What the World Thinks About Climate Change in 7 Charts," Pew Research Center, April 18, 2016; "U.S. Concern About Global Warming at Eight-Year High," Gallop Poll, March 2016.

105 "Science tells us the climate is changing": Dan Merica and Rene Marsh, "Trump's EPA Pick: Human Impact on Climate Change Needs More Debate," CNN online, January 18, 2017.

107 "The signature effects": Risky Business Project, "National Report: The Economic Risks of Climate Change in the United States," June 2014.

108 "frame climate change as an economic issue": Alicia Mundy, " 'Risky Business' Report Aims to Frame Climate Change as an Economic Issue," *Wall Street Journal*, June 23, 2016.

108 "John D. Rockefeller, the founder of Standard Oil": see Suzanne Goldenberg, "Heirs to Rockefeller Oil Fortune Divest from Fossil Fuels over Climate Change," *The Guardian*, September 22, 2014; see also Rockefeller Brothers Fund, Divestment Statement, August 31, 2016, www.rbf.org/about/divestment.

109 responded with a hard-hitting two-part essay: David Kaiser and Lee Wasserman, "The Rockefeller Family Fund vs. Exxon," *New York Review of Books*, December 8, 2016, and December 22, 2016.

111 The Trump transition team also became indirectly involved via the Breitbart News Network: John Schwartz, "Exxon Mobil Accuses the Rockefellers of a Climate Conspiracy," *New York Times*, November 21, 2016.

112 "lenders, insurers, and investors": "Our Mission," Task Force on Climate-related Financial Disclosures website, www.fsb-tcfd.org, 2016.

112 "Climate change is not only": Hiroko Tabuchi, "Tell Investors of Climate Risks, Energy Sector Is Urged," *New York Times*, December 14, 2016.

112 "stranded assets": Carbon Tracker Initiative, "Unburnable Carbon 2013: Wasted Capital and Stranded Assets," April 2013, www.carbon tracker.org/wp-content/uploads/2014/09/Unburnable-Carbon-2 -Web-Version.pdf.

114 took the form of human beings: see Samuel H. Williamson and Louis Cain, "Measuring Slavery in 2011 Dollars," MeasuringWorth.com; and Christopher Hayes, "The New Abolitionism," *The Nation*, May 12, 2014.

116 In April 2014 the CEO of ExxonMobil: ExxonMobil, "Energy and Carbon—Managing the Risks," available at http://cdn.exxon mobil.com/~/media/global/files/energy-and-environment/report ---energy-and-carbon---managing-the-risks.pdf; and "Exxon Sees Little Climate Change Risk to Assets," Reuters, March 31, 2014.

118 "When will humanity express its moral outrage": see Duane Elgin, "Global Warming and Carbon Dioxide Ethics," *Huffington Post* (blog), August 23, 2017.

119 Later "sea-level rise" was reinstated: Amanda Terkel, "Ban on 'Climate Change' Widespread Across Florida Government Under Rick Scott," *Huffington Post*, March 12, 2015.

119 "it will be hard to plan": "In Florida, Officials Ban Term 'Climate Change,'" Florida Center for Investigative Journalism website, March 8, 2015.

120 Scott would later deny . . . "I'm into solutions": see "Florida Governor Denies Environmental Agency Banned Term 'Climate Change,'" *Miami Herald*, May 9, 2017.

121 "This storm will kill you!": see Pamela Engel, "Florida Gov. Rick Scott on Hurricane Matthew," *Business Insider*, October 6, 2016.

121 Significant in the evolving awareness: see for example John Hocevar, "Hurricane Matthew Is an Ugly Example of What Happens When Climate Denial and Extreme Weather Collide," Greenpeace website, October 10, 2016.

122 Ten Florida scientists wrote to Donald Trump: see "Florida Scientists

Call for Climate Meeting with Trump," *Miami Herald*, December 22, 2016.

122 "If you're truly serious": "Governor Brown to Governor Scott: 'Time to Stop the Silly Political Stunts and Start Doings Something About Climate Change," Office of Governor Edmund G. Brown Jr., May 2, 2016.

122 Brown's state of California: "The Governor's Climate Change Pillars: 2030 Greenhouse Gas Reduction Goals," California Environmental Protection Agency, Air Resources Board, September 20, 2016.

122 "wind turbines that rise out of the cornfields": Jeff Biggers, "Cities and States Lead on Climate Change," *New York Times* (op-ed), November 30, 2016.

123 increase in the number of Republicans: Coral Davenport, "Conservative to Fund Republicans Who Back Climate Change Action," *New York Times*, June 29, 2016; and John Schwartz, "'A Conservative Climate Solution': Republican Group Calls for Carbon Tax," *New York Times*, February 8, 2017.

125 He acted on these convictions: see Nancy L. Rosenblum, ed. *Thoreau's Political Writings* (Cambridge: Cambridge University Press, 1996).

128 "sinister and little-recognized partners": Rachel Carson, *Silent Spring* (New York: Houghton Mifflin Harcourt [1962], 2002).

129 "two products of wartime science": Linda Lear, *Rachel Carson: Witness for Nature* (New York: Houghton Mifflin Harcourt [1997], 2009), 374.

129 "These [nuclear] pollutants . . . shapers of it": Ralph H. Lutts, "Chemical Fallout: Rachel Carson's Silent Spring, Radioactive Fallout, and the Environmental Movement," *Environmental Review* 9, no. 3 (Autumn 1985), 210–25, 211.

130 She was nonetheless attacked: Nancy F. Koehn, "From Calm Leadership, Lasting Change," *New York Times*, October 28, 2012.

130 contemporary groups who falsify: see for example Gordon J. Edwards, "The Lies of Rachel Carson," *21st Century Science & Technology*; and "Rachel Carson's Dangerous Legacy," SafeChemicalPolicy.org, March 1, 2007.

130 "didn't just deny the facts of science": Oreskes and Conway, *Merchants of Doubt*.

131 staggering statistic of 7 million: World Health Organization, news release, March 24, 2014.

133 "laws and governance": Polly Higgins, *Eradicating Ecocide* (London: Shepheard-Walwyn, 2010).

8: Climate Swerve 2: Awareness and Adaptation

139 "every person living on this planet. . . . common home": Pope Francis, "Laudato Si: On Care for our Common Home," encyclical letter, 2015.

140 treaty between China and the United States: *New York Times*, November 11, 2014.

140 ratification of the Paris agreement: Jean Chemnick, "U.S. and China Formally Commit to Paris Climate Accord," *Scientific American*, September 6, 2015.

145 "grim new low": Suzanne Goldenberg, "Arctic Sea Ice Crashes to Record Low for June," *The Guardian*, July 7, 2016.

147 demise of "the hope": Dale Jamieson, *Reason in a Dark Time* (New York: Oxford University Press, 2014).

151 *proteanism*: Robert Jay Lifton, *The Protean Self: Human Resilience in an Age of Fragmentation* (New York: Basic Books, 1995).

152 "transforms the whole": Ernst Cassirer, *An Essay on Man: An Introduction to a Philosophy of Human Culture* (New Haven, CT: Yale University Press [1944], 1962].

Index

Index

Bravo test, 24
Breitbart News Network, 111
Brown, Jerry, 122
Buber, Martin, 37
Buffalo Creek, West Virginia, 40–41
Bulletin of the Atomic Scientists, 27
Bush, George W., 110

California, 110, 122
carbon dioxide, 56–57
Carbon Dioxide: The Good News
   (Goklany), 56–57
carbonism, 56–57
Carbon Tracker Initiative, 112–13
Carey, Michael, 78
Carson, Rachel, 127–30
Cassirer, Ernst, 151, 152
"catastrophic environmentalism," 29–30
"Chemical Fallout" (Lutts), 129
Chernobyl nuclear accident, 58–59, 62
Chevron, 108, 126
China:
   climate change agreements of,
      140–41
   and Nuclear Nonproliferation Treaty,
      27
   nuclear testing in, 88
   thought reform in, 67–68, 95
Christian apocalyptic narratives, 37–38,
   41–43
chronos, 89–90
civil defense, 76–80
climate activism:
   and environmentalism, 124–25
   witnessing professionals in, 99
climate change:
   current evidence of, 53, 54
   denials of. *see* climate rejection
   economic consequences of, 107–12
   experiencing effects of, 104–7
   in Hiroshima, 26
   images of, 65–66
   increased media coverage of, 104–5
   in Marshall Islands, 25–26
   nuclear winter as, 23

psychic numbing for, 36
and species awareness, 155–56
Edward Teller on, 33
climate change rejection, 91. *See also*
   climate rejection
climate normality:
   and formed awareness, 136–37
   and malignant normality, 87–92
   professionals contributing to, 97
climate nuclearism, 57–58
climate rejection:
   changing strategies of, 105–7
   congressional actions supporting, 111
   consequences of, 119–24
   defining, xii
   by fossil fuel companies, 109–10
   and fossil fuel interests, 55
   in politics, 88–89
climate scientists:
   activism of, x, 55–56
   mental struggles of, 52–56
   as witnessing professionals, 100
climate swerve, 101–56
   and adaptation paradox, 148–50
   as call to action, 146–48
   and Rachel Carson, 127–30
   and death anxiety, 102–4
   and denials of climate change, 119–24
   and ecocide, 132–34
   economic indicators of, 107–12
   and environmentalism, 124–25
   experiencing effects of climate change
      in, 104–7
   and formed awareness, 135–37
   and fossil fuels as stranded assets,
      112–15
   grim indicators, 144–46
   and humans as prospective survivors,
      153–56
   and humans as single threatened
      species, 139–41
   pollution affecting health and illness
      in, 130–32
   and pollution awareness, 126–27
   and possibilities of proteanism, 150–53

Index

Index

Index

Index

nuclear tests:
  across the world, 87–88
  Alamogordo test, 51, 87–88
  Bravo test, 24
  as environmental experiments, 27–30
  in Marshall Islands, 23–24
  pollution from, 127–30
  radiation from, 128–29
  Trinity test, 48–49, 72
nuclear threats, 17–43
  and apocalyptic narratives, 37–43
  climate threats merging with, 30–32
  climate threats vs., 45
  dark predictions of, 144
  in Marshall Islands, 23–27
  and nuclearism, 32–35
  nuclear tests as environmental
    experiments, 27–30
  psychic numbing to, 35–37
  Earle Reynolds's response to, 18–23
nuclear weapons:
  and imagery of extinction, 17
  and malignant normality, 70–74
  and nuclear disarmament, 26–27
  and nuclear power plants, 58, 64–65
nuclear winter, 22–23
Nye, Joseph, 81

Obama, Barack, 131
On the Beach (Shute), 28–29
Oppenheimer, Robert, 48–52, 54
Oreskes, Naomi, 54, 55, 109–10, 130
Orwellian (term), 121
Oxford English Dictionary, 94

Pakistan, 27
Papal encyclical (2015), 139–40
Paris Climate Conference:
  and climate rejection, 66
  as demonstration of species awareness,
    xi, 144–45
  influence of earlier statements on,
    139–40
  and stranded adaptation, 137–38
Parsons, William S., 73–74

Paulson, Henry, 107
Phoenix (yacht), 19–20
Physicians for Social Responsibility
    (PSR), 141
Pickens, Slim, 83
Pilgrim nuclear power plant,
    Massachusetts, 62–63
politics, and climate rejection, 88–89, 120
pollution:
  affecting health and illness, 130–32
  awareness of, 126–27
  and health, 130–32
  and Hiroshima, 10–11
  nuclear radiation as, 29
positive nuclear scenarios, 71–74
professionals, 93–100
  defined, 94–95
  in military service, 95–96
  Nazi doctors, 96–97
  in Vietnam War, 93–94
  as witnessing professionals, 98–100
prophetic survivors, 50–52
prospective survivors:
  and dark predictions future, 146–48
  humans as, 153–56
proteanism, 150–53
Proteus, 151
Pruitt, Scott, 105
PSR (Physicians for Social
    Responsibility), 141
psychiatrists:
  on malignant normality, 74–76
  normality and diagnoses for, 68–69
psychic numbing:
  and climate normality, 92
  to climate threats, 99
  and fragmentary awareness, 136
  and nuclear normality, 79–80
  to nuclear threats, 35–37
psychism, 80
psychohistorical dislocation, 103, 151
psychological effects:
  of climate normality, 92
  of "duck and cover" drills, 78–80
  of nuclear accidents, 60–62

Index

Index

About the Author

Robert Jay Lifton has written more than twenty books and edited many others. His seminal works on psychology and history include the National Book Award–winning *Death in Life: Survivors of Hiroshima*, *The Nazi Doctors: Medical Killing and the Psychology of Genocide*, and *Witness to an Extreme Century: A Memoir*.

Celebrating 25 Years of Independent Publishing

Thank you for reading this book published by The New Press. The New Press is a nonprofit, public interest publisher celebrating its twenty-fifth anniversary in 2017. New Press books and authors play a crucial role in sparking conversations about the key political and social issues of our day.

We hope you enjoyed this book and that you will stay in touch with The New Press. Here are a few ways to stay up to date with our books, events, and the issues we cover:

- Sign up at www.thenewpress.com/subscribe to receive updates on New Press authors and issues and to be notified about local events
- Like us on Facebook: www.facebook.com/newpressbooks
- Follow us on Twitter: www.twitter.com/thenewpress

Please consider buying New Press books for yourself; for friends and family; or to donate to schools, libraries, community centers, prison libraries, and other organizations involved with the issues our authors write about.

The New Press is a 501(c)(3) nonprofit organization. You can also support our work with a tax-deductible gift by visiting www.thenewpress.com/donate.